Global Warming and Energy Policy

Global Warming and Energy Policy

Edited by

Behram N. Kursunoglu
Global Foundation, Inc.
Coral Gables, Florida

Stephan L. Mintz
Florida International University
Miami, Florida

and

Arnold Perlmutter
University of Miami
Coral Gables, Florida

Springer Science+Business Media, LLC

Library of Congress Cataloging-in-Publication Data

Global warming and energy policy/edited by Behram N. Kursunoglu, Stephan L. Mintz, and Arnold Perlmutter.

p. cm.

Includes bibliographical references and index.

ISBN 978-1-4613-5497-0 ISBN 978-1-4615-1323-0 (eBook)

DOI 10.1007/978-1-4615-1323-0

1. Global warming—Government policy—Congresses. 2. Energy policy—Congresses. I. Kursunoglu, Behram, 1922- II. Mintz, Stephan L. III. Perlmutter, Arnold, 1928-.

QC981.8.G56 G58145 2001

363.738'747—dc21

2001029946

Proceedings of an annual energy symposium, held November 26–28, 2000, in Fort Lauderdale, Florida

ISBN 978-1-4613-5497-0

Originally published by Kluwer Academic / Plenum Publishers, New York in 2001

Softcover reprint of the hardcover 1st edition 2001

10 9 8 7 6 5 4 3 2 1

A C.I.P. record for this book is available from the Library of Congress.

PREFACE

The first part of the conference explores two major environmental concerns that arise from fuel use: (1) the prospect that the globe will become warmer as a result of emissions of carbon dioxide, and (2) the effect upon health of the fine particles emitted as combustion products. The conference focused on the fact that there was lack of data direct enough to enable us to predict an entirely satisfactory result, and that makes policy options particularly difficult. With regard to (1) above, in the second half of the 20th century there were major increases in anthropogenic CO_2 emissions, and it is generally agreed that these were responsible for an increase in CO_2 concentrations. But the relationship between global temperature and CO_2 concentrations remains murky. The principal problem is that water vapor is a more important greenhouse gas than CO_2 and that the concentrations of water vapor vary widely in time and space.

The approach to this problem is probably, but not certainly, a positive feedback effect: as temperature increases so does the water vapor leading to further temperature increases. Scientists associated with the Intergovernmental Panel on Climate Change (IPCC) tend to believe the general features of the models. Other scientists are often less convinced. This conference had speakers that: (a) outlined the situation of predicting temperature rise, (b) outlined the present situation on the effect of temperature on economic activity, (c) discussed what steps can be taken to clarify this situation, and (d) what society might do while waiting for these steps to produce results. In regards to (2), the situation about the effects of air pollution on health is less global, although many parts of the world are affected. Direct epidemiology is hard put to attribute risks that are less than a doubling of health effects. The scientific uncertainties therefore remain almost as large as those for global warming.

While almost everyone would agree that the high exposure in the air pollution "incidents" of the past (such as in London in December 1952, which were directly experienced (by Kursunoglu and his wife) on the way to take the SS Queen Elizabeth to the United States) were clearly bad, there is disagreement of the effects of low exposures.

Behram N. Kursunoglu
Stephan L. Mintz
Arnold Perlmutter
Coral Gables, Florida
March 2001

GLOBAL FOUNDATION'S RECENT CONFERENCE PROCEEDINGS

Making the Market Right for the Efficient Use of Energy
Edited by: Behram N. Kursunoglu
Nova Science Publishers, Inc., New York, 1992

Unified Symmetry in the Small and in the Large
Edited by: Behram N. Kursunoglu, and Arnold Perlmutter
Nova Science Publishers, Inc., New York, 1993

Unified Symmetry in the Small and in the Large - 1
Edited by: Behram N. Kursunoglu, Stephen Mintz, and Arnold Perlmutter
Plenum Press, 1994

Unified Symmetry in the Small and in the Large - 2
Edited by: Behram N. Kursunoglu, Stephen Mintz, and Arnold Perlmutter
Plenum Press, 1995

Global Energy Demand in Transition: The New Role of Electricity
Edited by: Behram N. Kursunoglu, Stephen Mintz, and Arnold Perlmutter
Plenum Press, 1996

Economics and Politics of Energy
Edited by: Behram N. Kursunoglu, Stephen Mintz, and Arnold Perlmutter
Plenum Press, 1996

Neutrino Mass, Dark Matter, Gravitational Waves, Condensation of Atoms and Monopoles, Light Cone Quantization
Edited by: Behram N. Kursunoglu, Stephen Mintz, and Arnold Perlmutter
Plenum Press, 1996

Technology for the Global Economic, Environmental Survival and Prosperity
Edited by: Behram N. Kursunoglu, Stephen Mintz, and Arnold Perlmutter
Plenum Press, 1997

25th Coral Gables Conference on High Energy Physics and Cosmology

Edited by: Behram N. Kursunoglu, Stephen Mintz, and Arnold Perlmutter
Plenum Press, 1997

Environment and Nuclear Energy
Edited by: Behram N. Kursunoglu, Stephan Mintz, and Arnold Perlmutter
Plenum Press, 1998

Physics of Mass
Edited by: Behram N. Kursunoglu, Stephan Mintz, and Arnold Perlmutter
Plenum Press, 1999

Preparing the Ground for Renewal of Nuclear Power
Edited by: Behram N. Kursunoglu, Stephan Mintz, and Arnold Perlmutter
Plenum Press, 1999

Confluence of Cosmology, Massive Neutrinos, Elementary Particles & Gravitation
Edited by: Behram N. Kursunoglu, Stephan Mintz, and Arnold Perlmutter
Plenum Press, 1999

International Energy Forum 1999
Edited by: Behram N. Kursunoglu, Stephan Mintz and Arnold Perlmutter
Plenum Press, 2000

International Conference on Orbis Scientiae 1999
Quantum Gravity, Generalized Theory of Gravitation and Superstring Theory-based Unification
Edited by: Behram N. Kursunoglu, Stephan Mintz, and Arnold Perlmutter
Plenum Press, 2000

Global Warming and Energy Policy 2000
Edited by: Behram N. Kursunoglu, Stephan Mintz and Arnold Perlmutter
Plenum Press, 2001

International Conference on Orbis Scientiae 2000
The Role of Attractive and Repulsive Gravitational Forces in Cosmic Acceleration of Particles; The Origin of the Cosmic Gamma Rays
Edited by: Behram N. Kursunoglu, Stephan Mintz and Arnold Perlmutter
Plenum Press, 2001

Global Foundation, Inc.

A Nonprofit Organization for Global Issues Requiring Global Solutions, and forProblems on the Frontiers of Science

Center for Theoretical Studies

International Conference on

"Global Warming and Energy Policy"

(23rd In A Series of Conferences Since 1974)

November 26 - 28, 2000

Lago Mar Resort
Fort Lauderdale, Florida

Meeting
Lakeview Room

Sponsored by:
Global Foundation Inc.

P. O. Box 249055
Coral Gables, Florida 33124-9055
Phone: (305) 669-9411
Fax: (305) 669-9464
E-mail:

Conference Hotel:
Lago Mar Resort

1700 South Ocean Lane
Fort Lauderdale, Florida 33316
Ph: 1-800-255-5246
Fax: (954) 524-6627
Deadline: October 15, 2000
Group rate: $125.00/night

TOPICS WILL INCLUDE:

1. A Scientific Assessment Of Emission Of Greenhouse Gases Into The Atmosphere.

2. The Impact Of Preventive Measures To Arrest The Growth Of Global Warming On The Economy.

3. The Known Facts On The Contribution To Any Warming Effects By All Transportation Means Energized By Fossil Fuels.

4. When Can We Believe In The Validity Of Our Scientifically Based Judgment On The Existence Of A Global Warming?

5. The Value Of Education On Environment And Its Role In The Scientific Assessment Of Global Warming And Needed Preventive Affordable Actions.

Précis

The first part of the conference will explore two major environmental concerns that arise from fuel use: (1) the prospect that the globe will become warmer as a result of emissions of carbon dioxide, and (2) the effect upon health of the fine particles emitted as combustion products. For neither of these are the data direct enough to enable entirely satisfactory prediction, and that makes policy options particularly difficult. With regard to (1) above, in the second half of the 20th century there were major increases in anthropogenic CO_2 emissions, and it is generally agreed that these were responsible for an increase in CO_2 concentrations. But the relationship between global temperature and CO_2 concentrations remains murky. The principal problem is that water vapor is a more important greenhouse gas than CO_2 and that the concentrations of water vapor vary widely in time and space.

There is probably, but not certainly, a positive feedback effect: as temperature increases so does the water vapor leading to further temperature increases. Scientists associated with the Intergovernmental Panel on Climate Change (IPCC) tend to believe the general features of the models. Other scientists are often less convinced. This conference will have speakers to: (a) outline the situation of predicting temperature rise. (b) outline the present situation on the effect of temperature on economic activity. (c) discuss what steps can be taken to clarify this situation and (d) what society might do while waiting for these steps to produce results. In regards to (2), the situation about the effects of air pollution on health is less global, although many parts of the world are affected. Direct epidemiology is hard put to attribute risks that are less than a doubling of health effects. The scientific uncertainties therefore remain almost as large as those for global warming.

While almost everyone would agree that the high exposure in the air pollution "incidents" of the past (such as in London in December 1952, which were directly experienced by Dr. Kursunoglu and his wife on the way to take the SS Queen Elizabeth to the United States, it was this most unusual atmospheric phenomena that induced the British to stop burning coal) were clearly bad, there is disagreement of the effects of low exposures. This conference will study (A) the status of the science (B) what simple remedial steps might be available (C) steps to improve

the understanding. For this part of the conference we recommend Richard Wilson 's book on "Particles in Our Air"

The second part of the conference will comprise of local and regional effects, exhaustion of resources, geopolitical uncertainty on the production of electricity and the choice of energy sources in particular the future role of nuclear energy in the reduction of prevention of Global Warming.

--NOTES--

1. **Each presentation is allotted a maximum of 30 minutes and an additional 5 minutes for questions and answers.**

2. **Moderators are requested not to exceed the time allotted for their sessions.**

Moderator: Presides over a session. Delivers a paper in own session, if desired, or makes general opening remarks.

Dissertator: Presents a paper and submits it for publication in the conference proceedings at the conclusion of the conference.

Annotator: Comments on the dissertator's presentation or asks questions about it upon invitation by the moderator.

CONFERENCE PROCEEDINGS

1. **The conference portfolio given to you at registration contains instructions to the authors from the publisher for preparing typescripts for the conference proceedings.**

2. **Papers must be received at the Global Foundation by January 8, 2001.**

An edited Conference Proceedings will be submitted to the Publisher by February 15, 2001.

GLOBAL WARMING AND ENERGY POLICY 2000
PROGRAM

Sunday, November 26, 2000

8:00 - 12:00 PM **Registration**

9:00 - 11:30 AM **International Advisory Committee Meeting,** Director's Room

12:00 PM **Committee Members Lunch,** Director's Room

1:00 PM Opening Session: **Behram N. Kursunoglu**
Chairman of the Global Foundation, Inc.
Opening Remarks

Edward Teller, Lawrence Livermore Laboratory
and
Behram N. Kursunoglu
"Global Warming & Energy Policy"

Keynote Address: **Richard Wilson,** Harvard University
"CO_2 and particulates: large Tran boundary Pollution Problems, What is the Science? What are the Uncertainties? What action should we recommend?"

2:30 PM Coffee Break

3:00 PM SESSION I: **A Scientific Assessment Of Emission Of Greenhouse Gases Into The Atmosphere**

Moderator: **Jean Couture,** Former Energy Secretary of France

Dissertators: **Richard Wilson,** Harvard University
"Economics, Politics, and Science of Global Warming"

Klaus S. Lackner, Los Alamos National Laboratory
"Strategies for A Zero Emission Economy"

Randell Mills, CEO of Blacklight Power, Princeton, N.J.

"Blacklight Power Technology—A New Clean Energy Source with Potential for Direct Conversion to Electricity"

Keith Paulson, Westinghouse
"Nuclear Power—Meeting the Future Electrical Generation Paradigms"
Jean Couture, Paris, France
"The Precaution Principle: A Guide for Action"

Annotator: **Peter Beck**, Royal Institute of International Affaires

Session Organizer: **Behram N. Kursunoglu**

5:30 PM SESSION II: **Exhaustion of Resources, Geopolitical Uncertainties**

Moderator: **C. Pierre Zaleski**, Université Paris Dauphine

Dissertators: **Bertrand Vieillard-Baron,** Franatome, Paris, France
"The External Cost of Global Warming and International Carbon Prices"

Pierre Bauquis, Special Advisor, Chairman of Total Fina
"Constraints on fossil fuel's supplies for next half century"

Jacques Maire, President of IFE
"Gas resources for the 21st century"

Annotator: **Keith Paulson**, Westinghouse

Session Organizer: **C. Pierre Zaleski**

7:00 PM *Conference adjourns for the day*

Monday, November 27, 2000

8:30 AM SESSION III: **Nuclear Energy and Environment: Facts and Myths**

Moderator: **Bertram Wolfe, GE Nuclear**

Dissertators: **Leonard L. Bennet and C. Pierre Zaleski,**
Universite Paris Dauphine
"Nuclear Energy scenarios in the 21st century; potential for alleviating greenhouse gas emissions and conserving fossil fuels"

Phillippe Savelli, OECD/NEA, Paris
"The Role of Nuclear Energy in Sustainable Resource Management"

Shelby T. Brewer, Commodore Nuclear, Inc.
"Nuclear Plant Financial Performance in a Restructured Utility System"

Annotators: **Myron Kratzer**, Bonita Springs, Florida

Session Organizer: **Bertram Wolfe**

10:00 AM Coffee Break

10:30 AM SESSION IV: **Nuclear Energy and Non-Proliferation**

Moderator: **Edward Arthur**, Los Alamos National Laboratory

Dissertators: **Edward Arthur**, Los Alamos National Laboratory
"Emerging Issues Problems in the Proliferation Problem"

Myron Kratzer, Bonita Springs, Florida
"Nonproliferation Safeguards: To see or not to see"

J. Magill R. Schenkel, Institute of Transuranic Elements
"Non-Proliferation Issues for Generation 4 Power Systems: Advanced Waste Management"

Annotator: **Behram N. Kursunoglu**, Global Foundation, Inc.

Session Organizer: **Edward Arthur**

12:00 Noon: **Lunch Break**

1:30 PM SESSION V: **The Future of Nuclear Energy and Environment**

Moderator: **Angie Howard**, Nuclear Energy Institute

Dissertators: **Peter Beck**, Royal Institute of International Affaires
"Global Warming—An Opportunity for Nuclear? Well, Yes, But…"

Carl E. Walter, Lawrence Livermore Laboratory
"Near-term Demonstration of Benign, Sustainable, Nuclear Power"

Angie Howard, Nuclear Energy Institute
"The Future of Nuclear Energy in the U.S.A."

Annotators: **Gerald Magill**, Saint Louis University

Session Organizer: **Angie Howard**

3:00 PM Coffee Break

3:30 P.M. SESSION VI: **Update and Overview of Global Warming**

Moderator: **Mike MacCracken,** National Assessment Coordination

Dissertators: **Mike MacCracken**
"Global Warming: A Science Overview"

Domenico Rossetti di Valdalbero, European Commission
"Energy technologies and climate change a world and European outlook"

Annotator: **Thomas Cunnington,** Cyton Corporation

Session Organizer: **Mike MacCracken**

5:00 PM *Conference adjourns for the day*

6:30 PM **Cocktail Reception, Fountainview Lobby**
Courtesy of Lago Mar Resort & Club

7:30 PM **Conference Banquet, Tidewater Room**

Tuesday, November 28, 2000

8:30 AM SESSION VII: **Reliability of Data and its Impact on Prediction Pertaining to Global Warming**

Moderator: **Gerald Clark,** Uranium Institute, London

Dissertator: **Juan Eibenschutz,** Luz y Centro ,Mexico
"Is Nuclear Energy Going to Miss Its Environmental Mission?"

Annotator: **Arnold Perlmutter,** University of Miami

Session Organizer: **Gerald Clark**

10:00 AM Coffee Break

10:30 AM **Concluding Panel Discussion**

Moderator: **Richard Wilson**

Panel Members: **Gerald Clark,** Uranium Institute London

Jean Couture, Paris, France

Mike MacCracken, National Assessment Coordination Office

C. Pierre Zaleski, Université Paris Dauphine

12:30 PM ***Global Foundation's 2000 Energy Conference Adjourns***

CONTENTS

INTRODUCTION

A SCIENTIFIC ASSESSMENT OF EMISSION OF GREENHOUSE GASES INTO THE ATMOSPHERE

NUCLEAR ENERGY AND ENVIRONMENT: FACTS AND MYTHS

FIVE ISSUES

Dr. Behram N. Kursunoglu, Dr. Edward Teller*

In the following, five important issues shall be discussed. We believe that their proper handling will greatly contribute to world stability.

We want to emphasize that this paper proceeds along lines which are unusual in science. In science, it is general practice to solve basic problems first, and then to proceed to fill in details.

In the following, five details will be considered: energy, food, climate, war and science. The big basic problems are world government and avoidance of overpopulation. In our opinion, these are at present too difficult to solve, and at the same time, it is much too tempting to accept inadequate solutions. A solution or even an incomplete solution of the five special problems mentioned above will make it easier to attack the bigger problems. Furthermore, the solution of the five problems mentioned contains some scientific and unusual elements. This makes their separate discussion profitable.

Making energy generally available consists primarily of the generation of electricity. As practiced today, this generation runs into two problems.

In the not-too-distant future, there will be a worldwide shortage of some energy sources from which electricity is generated. The present estimated reserves of fossil fuel include one trillion barrels of oil, four thousand trillion cubic feet of natural gas, large reserves of coal in China, Russia and U.S.A., with an equivalent energy content considerably greater than the sum of oil and natural gas reserves.

It is probably possible to calculate total emission of greenhouse gases into the atmosphere whereby generating a global warming debt might defeat the usefulness of the above-mentioned reserves. Therefore, we must look for energy sources without any environmental impact. Fission reactor is one good example with which we have become quite familiar, and it is likely that soon the 21st century economy may have to be based on fission.

The most widespread worry is that use of coal contributes to "greenhouse gasses" which tend to retain energy in the atmosphere and is apt to cause a rise in the average temperature. An observed rise of one-third of a degree Kelvin in fifteen years, is not

* Dr. Behram N. Kursunoglu, Global Foundation, Inc, Coral Gables, Florida 33124 USA. Dr. Edward Teller, Lawrence Livermore National Laboratory, Livermore, California 94551 USA, and Hoover Institution/Stanford University, Stanford, California 94305-6010 USA.

Global Warming and Energy Policy, Edited by Kursunoglu *et al.*
Kluwer Academic/Plenum Publishers, New York 2001

certain enough to have real predictive value. What is more important, the impending shortage of fuel makes the proposed expensive solutions (by taxation) seem less necessary. But the impending shortage of some fossil fuels does make it desirable to look for another energy source. The obvious choice of such an energy source is energy from nuclear reactors. Actually, one big nuclear reactor accident in Chernobyl in 1986 made nuclear reactors unpopular, and they appear to be on the decline in some parts of the world. This is a mistake, in our opinion. The decline should be stopped and reversed. A radical change such as replacing nuclear fission (splitting of big nuclei) by nuclear fusion (fusing small nuclei) seems too difficult and unnecessary. Indeed, explosive energy releases based on fusion are well advanced, while plans for a machine continuing to deliver fusion energy appear much less promising.†

On the other hand, worries about accidents in fission reactors appear exaggerated. In the half-century in which these reactors have been working in several parts of the world, there have been only three major accidents. There was one in England, Windscale in 1956; Three Mile Island in Pennsylvania in 1979; and the third in the Soviet Union in 1986 in Chernobyl. The damage in each of these was estimated in the billion-dollar category, so that incentive for future safety is assured. In the first two, no one seems to have been hurt. In Chernobyl, not much less than 100 were killed as an immediate consequence, including the fighting of the fire that ensued. The radioactive fallout from that accident could not be confined and reached as far away as Turkey. There have been claims of tens of thousands of people having been seriously hurt by this ejected radioactivity, but a conference in Vienna on the ten-year anniversary found that actually the only large-scale damage was due to fear.‡

It is reasonable to compare the number of casualties divided by the amount of electricity generated. This number is less for fission reactors than for other widespread energy generators such as those burning coal. It might, therefore, make sense to plan future energy generators to be based on nuclear fission. We should also like to suggest that fear of radioactivity should be diminished by effectively eliminating dispersion of radioactivity from fission reactors. This could be accomplished by locating nuclear reactors in loose dry earth a thousand feet underground and by constructing them in such a way that they should have a negative temperature coefficient. (This new concept of underground placement of energy generating reactors needs an economic assessment in particular, in case of an accident, how that scenario compares to reactors operating above the ground. In this case, we need to do two things: 1) Discussion of the projected reactor house; 2) The structure and waste management—Should it be preserved where the waste is produced or be transported somewhere else? Both cases have pros and cons.) In order to reduce the danger, the construction should be such that if reactor temperature increases, its energy output should decrease, and that at a moderately increased temperature, the reactor should stop producing energy. To do this, we should rely on the contribution of slow neutrons to the reactions, and also add an appropriate amount of

† Reasonable cost predictions remain high in spite of efforts in Russia and the United States on improvements in the planning of future fusion reactors.

‡ In the following months, the number of abortions in Western Europe showed an excess of a few times 10,000, due to unjustified fears that the offspring could suffer from the radioactive fallout generated by Chernobyl.

neutron absorbers, which, due to a resonance, absorb neutrons more effectively in the epithermal than in the thermal regions.

The reactor should be cooled by the inert gas, helium, and could be regulated by the flow of helium. If more helium is pumped, the reactor cools down, and its energy output increases. If too much energy is produced, the reactor heats up and the energy production declines. Thus, the reactor cannot overproduce energy by overheating, and substantial energy production would go on only in a vigorously cooled state.

The generation of electricity would be a two-stage process. One stage, a thousand feet underground, would produce not electricity, but hot helium; the second stage at very shallow depth would transfer the high energy of helium into electricity by a conventional generator, which, being at shallow depth, is easily accessible.

There have been numerous experimental underground nuclear explosions performed in Nevada. It was found that after the explosion, the produced substances are practically immobile in the absence of water. The reactor here described is inherently stable, but even in case of an accident, the radioactivity will stay underground if the earth is dry. The radioactivity from an exploded underground reactor should actually not worry us because resulting heat from radioactivity evaporates any invading water and keeps the residue dry. Thus we have double safety. No uncontrolled reaction or "run-away" could occur because of the negative temperature coefficient, and if such a run-away actually should nevertheless take place, then there will be a lack of means to transport the radioactivity to the surface of the earth.

The functioning of the reactor and the investment in the reactor will not be at risk except through gross errors in the construction or through a most improbable reconfiguration of the underground rock by earthquake or meteor impact. But even in that case, there will be no health damage because the radioactivity will remain confined. Ultimately, if the radioactivity is to reach the surface, the primary cause (earthquake, volcanic flow, or meteorite impact) must itself occur all the way between the surface and the depth of 1,000 feet, which amounts to total destruction due to the external cause rather than the reactor.

Had we located the reactor in solid rock, a splitting of the rock by an earthquake could have brought radioactivity to the surface. By placing the reactor in loose dry earth, an earthquake will leave the earth loose and the radioactivity immobile.

The whole energy-producing process is to be started by a conventional uranium reactor that functions for one or a few years. It produces fission and energy and also an excess number of neutrons. The latter could be used to activate a thorium reactor which is initially incapable, by itself, of maintaining a reaction, but if activated by neutrons, a burning would start as follows: Th^{232} + neutron yields Th^{233}. Th^{233} yields beta + beta + U^{233}. U^{233} + neutron yields fission + neutrons. Th^{232} + neutron yields Th^{233} (as above) and the cycle starts again. Or, alternatively, we could continue to use U^{238} + neutron yields U^{239}. U^{239} yields beta + beta + Pu^{239}. Pu^{239} + neutron yields fission + neutrons. U^{238} + neutron yields U^{239}, and start again.

Thus, both the Th-based and the U-based cycles will work for a second reactor. But to initiate the Th-based cycle, some previous activation by neutrons is necessary, whereas in the case of uranium, the presence of some (fissionable) U^{235} helps in getting the process started. Thus, a breeder as such reactors are called, in thorium, requires a start-up by neutrons from a uranium reactor. The second thorium reactor may activate a third thorium reactor. This could continue in a chain of reactors for a millennium if we so choose. The point in using thorium is that it is much more abundant than uranium and will not be exhausted in 1,000 years.

According to plans, energy production by a reactor and its eventual shutdown should be followed by transporting residual radioactivity to a safe place, for instance, in Nevada. It seems to us that this is the most dangerous phase. A traffic accident in transportation might have catastrophic consequences. It seems to us simpler, less expensive, and, above all, more safe to leave the radioactivity where it has been produced, provided that the original location and configuration have been chosen with care. Some activity may remain significant for a millennium, but it will not be harmful, and eventually safe and profitable applications may develop for radioactivity; the exhausted reactor may be "mined" for its radioactive content.

A few additional remarks are needed. One is that the planned reactors require expert knowledge and careful execution in their original placement. But the subsequent operation, for possibly a considerable number of years, needs little expertise. Thus the use of reactors by the underdeveloped part of the world need not run into any difficulty.

The second point is that only the installation of the reactors will be expensive. Their operation will require little expertise (as stated above) and little expense. The result may be that nuclear energy may become truly competitive.

The number of thorium reactors taking over from a single uranium reactor need not be specified at the beginning. In a chain of reactors, only one reactor (i.e., uranium reactor or thorium reactor) will be active at one time except for the intervals where the reaction is handed over from a reactor to the next. We may guess that the operating cost for one (or two) reactors and the subsequent transformation into electrical energy will have an operating cost of not more than $10 million a year for an output of a thousand megawatts. That would mean $10 per kilowatt year and a couple of cents per kilowatt-hour.

The last figure, of course, has to be multiplied by at least a factor of five since the cost of distribution and the really big cost of building the reactors, have to be added. The competitive nature of the enterprise must ultimately depend on an effective, inexpensive and well-planned way of building the reactors. All we have done here is to indicate how low operating costs have to become in order to make the worldwide application of nuclear reactors feasible.

The third and most important point is that radical military misuse can be eliminated if the reactors are inaccessible (except by dangerous and expensive mining operations). Thus, the planned reactor will not make material available for military stockpiles.[§] (It appears that around the reactor, to avoid its radical military misuse, we can construct some kind of arrangement making access more difficult or at least more obvious.)

We claim that nuclear reactors are clearly the best possibilities for energy production. Other options, like solar energy, hydroelectric energy or geothermal energy, might be applicable for particular local uses, but for big-scale production, nuclear energy is apt to win. And the approach using fission seems to be the easy and inexpensive one.

[§] The residues might also be used for their plutonium content for military purposes. This danger will increase with the passage of time; but time also may permit the development of appropriate agreements and safeguards.

The one necessary limitation is the question of food supply. We see two possible sources for much more food. One is the exploitation of the oceans. Today's fishing is, in principle, not very different from Paleolithic hunting. When shall we cultivate the oceans?

There is a different approach for more food supply based on our rapid progress in the science of biology. The earliest human civilization made a great impact through the domestication of animals and plants. This required detailed knowledge and study of the mature animals or plants as well as of the connection between parent and offspring. Today, great advantages can be derived from our understanding of inheritance which functions through information carried by one type of organic molecule, deoxyribonucleic acid (DNA). Actually, the most impressive international study of the millions of components of the DNA molecule in humans has been recently completed. The same thing for other forms of life is underway in an impressive manner.

Unfortunately, there are widespread fears associated with newly-bred forms of food supplies because of their conceivably harmful nature. Are scientific developments really more dangerous than the Paleolithic development through trial and error? Need we be really afraid of knowledge and its consequences? It is a reasonable step to study of modifications of DNA (which proceeds spontaneously) and the elimination of undesirable modifications. What Paleolithic man accomplished in millennia, we might perform in a single generation through modern methods of inducing variations, observing the results in the animal or vegetable DNA and its consequences in the resulting organisms.

Environmentalists' objections have been remarkable in connection with possible effects human activities may have on the average temperature of the earth. That changes in climate may have catastrophic consequences is obvious. Indeed, in the last million years, there have been several ice ages. The trend now seems to be reversed and average temperatures appear to be slowly rising. This effect is considered by many as harmful and is attributed to "greenhouse gasses," principally CO_2, generated by the burning of coal.

A possible remedial approach would consist of a two-stage program. First, it would seem reasonable to initiate a big-scale effort on weather prediction. Weather prediction could be extended from only a five-day prediction at present to a two-week prediction to be reached within a few years. This would lead to savings of more than one billion dollars per decade in the United States alone and even more worldwide.** The second change we should recommend is to explore the possibility of weather modification, a possibility of great potential benefits, including prevention of the claimed worldwide rise in temperature if indeed that proves to be a real danger.

Let's consider first the practice of weather prediction. Weather prediction runs into the difficulty that future weather conditions depend sensitively on small changes at the present time. This phenomenon is encountered in many computational attempts where success has been frustrated by small causes giving rise to big consequences. Indeed, weather phenomena would appear to be ideal examples of calculable predictions by modern computing equipment except for the requirement of accurate initial conditions.

** Three main profits would be: By predicting the best times for planting and harvesting; by adapting long-range airplane flights to predicted wind velocities; and third, by saving lives and property from predicted floods and hurricanes.

It has been therefore proposed to introduce an atmospheric observational system consisting of two components: Approximately a billion few-inch size, thin-walled plastic spheres[††] at various altitudes all over the atmosphere which might stay afloat for an average of a few months; and secondly, approximately one hundred satellites at an altitude of a few hundred miles, so that they can readily interrogate the spheres floating in the underlying atmosphere. Short, laser pulses from these satellites would be reflected by the small plastic spheres by mounting on the latter a small apparatus called a corner reflector, which returns the reflected radiation into the direction from which it came. The time elapsed between emission and eventual return of the laser light pulses at each satellite will determine the distance of the reflecting objects, while the differential reflectivities will determine the local temperature and humidity at each reflector. One can keep track of the billion small atmosphere-probing spheres with the help of a hundred small satellites, and the data from the satellites also determine the positions and velocities of all of these small transponder objects. Perhaps, the biggest difficulty in execution (about which we are optimistic) is to mass-produce the small transponder-spheres for as little as 10 cents apiece and the satellites for $10M apiece. We expect that the transponder-spheres will be replaced a few times each year and the satellites once every ten years. Thus, the total cost of operating this atmospheric probing system will be approximately $1B per year. Our estimates show that such a system would be good enough to extend weather predictions to two weeks.

Such impressive savings could be further outdone by actually improving the weather. But, weather modification at once runs into the problem of not being to everyone's benefit, or at least of not being perceived as serving everyone's benefit. Yet, if one has arrived at an opinion that burning of coal modifies the weather, it appears to be an obvious necessity to consider other, perhaps compensating, influences. There is, indeed, a straightforward answer to the rising temperature. Distribution of particles of microscopic size in the stratosphere will scatter sunlight back into space and lower the temperature of the earth. Indeed, this effect has been observed in connection with volcanic eruptions.

We believe that it is premature to counteract an expected rise in temperature by distributing exceedingly small scatterers of sunlight in the stratosphere. We do believe, however, that relatively simple experiments should be carried out on small scatterers so that if need arises, such scatterers will be available for appropriate use.

The result should be to lower the temperature of the whole earth. While such a change might be approved by the majority, it may well be objected to by a minority. It seems, to us, necessary to carry out sufficient experiments to predict the effect and then to obtain consent and descent on the basis of predicting the actual changes that will be obtained. It is, of course, impossible to predict what the final decision should be in each case.

Explosives based on nuclear fission used over Hiroshima and Nagasaki ended the Second World War. This fact, together with the development of the hydrogen bomb, has given rise to widespread apprehensions about the damage that will accompany the fighting of a third world war. In many ways, this should be counted only as one more manner in which advancing technology has contributed to the horrors of war. Tanks,

[††] Filled with an appropriate mixture of helium and nitrogen.

airplanes and long-range missiles may make the third world war insupportable even in the absence of nuclear explosives.

There appears to be, however, one circumstance that makes nuclear explosives in themselves very specifically dangerous. That is the element of surprise. Knowledge of nuclear and hydrogen explosives has spread and will in the not too distant future be generally available. In this way, even a relatively small and not highly developed country may inflict truly horrible damage practically anywhere in the world. In other words, the unpredictable nature by which a nuclear war can be started is a qualitatively new element that should deeply worry everyone. We shall discuss a possible countermeasure below, but will first emphasize the dimensions of the problem.

The Pu^{239} that accumulates in fast reactors is used as fuel in slow reactors. It can also be collected in used fuel elements of the slow reactors. Unfortunately, Pu^{239} can also be used as an essential part of nuclear explosives. How does one prevent plutonium theft? The concern is that plutonium can be stolen from the nuclear-energy industry by terrorists for the construction of nuclear bombs. Vast literature exists on this subject discussing large numbers of scenarios of plutonium theft or laser-induced U^{235} extraction from natural uranium. The possibilities cannot be ignored, even though it is quite difficult to assign probabilities of occurrence to nuclear terrorist activities. Such security issues constitute an important part of the impact of nuclear power.

The problem of security for peaceful uses of nuclear energy was discussed in 1973 by Theodore B. Taylor, a theoretical physicist and former bomb designer, in a series of articles in The New Yorker Magazine. In a 1974 book, he gave information on how to make nuclear bombs. His efforts and the resulting publicity have led to better and greatly tightened safeguard procedures. None of these, even if the U.S. and Russia were the only nuclear monopolies, could be one hundred percent safe with regard to plutonium theft.

At present, a laser-based method exists to produce highly enriched uranium where the U^{235} isotope constitutes 90% of the uranium instead of the naturally occurring amount (less than one percent). Highly enriched uranium is strictly for bomb use and is independent of the nuclear power industry. Worldwide use of nuclear energy for electricity production presents other complications with regard to safeguarding fissionable fuel. The usefulness and safety of nuclear energy depend very much on international cooperation in all phases of its peaceful uses. The International Atomic Energy Agency, headquartered in Vienna, Austria, must be strengthened to emphasize the necessity of international cooperation for the safe use of nuclear energy. International cooperation is one of the most important moves toward making the world safe in the nuclear age. Failure or error in the nuclear age could affect the entire world. All of these difficulties are greatly increased by modern emphasis on secrecy. We believe that the introduction of an appropriate amount of openness will help to counteract the great dangers mentioned. We shall return to that important point at the end of our paper.

The danger of nuclear war should be specifically discussed in connection with the United States. Oceans and the general peaceful condition of the Western Hemisphere have provided a limitation to surprises to the United States. This historic advantage cannot last.

The situation was similar in England in the beginning of the 1930s. The British Isles had lost their protection from Europe. This fact played a considerable part in the political development prior to World War II.

One must conclude that the world at large and the United States in particular have a great stake in the establishment of missile defense. Somewhat more generally, one must acknowledge that in the age of military nuclear forces, all nations of the world face a particularly unstable future.

It is our conviction that the development of technology could lead to a better life for everyone. It would be highly desirable if that were true, and particularly if it would apply to stabilize behavior between nations. It is therefore a question of specific importance to diminish and if possible, eliminate the danger of surprise attack by nuclear means. There can be no doubt that this is a difficult question. We claim it is not unsolvable—indeed, we claim that, in the 1970s and 1980s, the U.S. had been on the right track toward a solution, which unfortunately has been abandoned.

The answer is missile defense executed a short time after the missile take-off. At that time, high accelerations are needed which make the take-off much more easily noticed. At that time also, there is less doubt as to responsibility for the launch. Finally, at that time, the intended victim may be uncertain. This last point, however, may be considered an advantage. Indeed, the defensive action is for protection of everyone. The point is, it is for the protection of any city and, therefore, serves the safety of the whole world instead of the safety of any one nation.

Earth-orbiting satellites have to serve to observe the preparation for the launch as well as the launch itself. Furthermore, the satellite should carry the needed counter-missile missiles. Using the rapidly improving methods of observation and the prediction of orbits, the counter-missile missiles could destroy its target by collision or by near-by conventional explosion or even, if need be, by a nuclear explosion.

But how do we know that the purpose of a missile is attack? Our suggestion is that we should assume that in all cases this is so except if the launching of a missile with date, orbit and detailed purpose has been announced appropriately ahead of time. Indeed, the ideal situation would be not to eliminate the launching of missiles but to limit them to internationally cooperative enterprises such as, for instance, weather observation as suggested previously.

It is unnecessary to reiterate the magnificent technical accomplishments that resulted from the science of the Twentieth Century. The previous section contained an appeal to use these same developments for the stabilization of the world. Unfortunately, there is, in contemporary science itself, a negative element that interferes with the realization of the dreams outlined above. That element is secrecy, which, of course, is in itself incompatible with one of the oldest and most important practices of science.

After the Oppenheimer hearings of 1954, Niels Bohr raised a strong protest against the fact that Oppenheimer lost his clearance. Bohr had also a long talk with one of the authors [ET], arguing with all his elaborate convictions against secrecy. He was so convinced of his argument that he did not notice that Edward Teller agreed with him.

As has been said earlier, the essential facts on the hydrogen bomb are practically no longer secret. Unfortunately, we see no way how to end secrecy by sudden and complete action. We do advocate abolishment of secrecy and the re-establishment of openness in science in a thoughtful and gradual manner.

What can be explained simply and clearly in one page cannot be kept secret. Details are less interesting from a general point of view and important only if you want sustained international cooperation. These are also the principles according to which the practical rules of private companies are established.

One point should be particularly emphasized—the simple things that need to be explained can be stated in their essence in one page each. The relativity of time as discovered by Einstein, the unpredictability of the future as discussed by Heisenberg, the release of nuclear energy as demonstrated over Hiroshima, the nature of the energy sources of the universe, all have been disclosed though perhaps not enough emphasis was placed on needed simplicity.

Our general recommendation, as has already been hinted, is not complete openness, but yet sufficient openness that can serve worldwide progress of science and worldwide establishment of rules.

We hope that applications of science as described in earlier sections, including cooperation between nations, will make it easier to approach a final necessary stage. That stage, as we see it, is not uniformity as might be enforced by world government, but maintenance of the differences between nations, which we consider to be a truly attractive part of man's activities. Such differences must be accompanied by patterns of behavior, which will tend to support cooperation and rule out the ultimately destructive function of war.

Therefore, this paper does not offer a solution but only a postponement of a great problem, together with hopes for a continually improving future.

ACKNOWLEDGEMENT

This work was performed in part under the auspices of the U.S. Department of Energy by the University of California Lawrence Livermore National Laboratory under Contract No. W-7405-Eng-48.

SOME GLOBAL ENVIRONMENTAL ISSUES OF PUBLIC CONCERN

Richard Wilson*

ABSTRACT

In this talk I will pick three key environmental issues that dominate the risk or impact, or dominate the public perception thereof:

(a) What is the effect upon health of particulate air pollution at today's levels? Experts increasingly believe that fine particulates kill 70,000 people a year in the USA, but this has not yet been officially admitted by any government.

(b) What is the effect of increased carbon dioxide emissions, from burning fossil fuels, on global climate change? There was intense government attention at Rio de Janeiro in 1992, in Berlin in 1995 and in Kyoto in 1996. The estimates of the International Program on Climate Change (IPCC) have changed but remain disturbingly high. What should the world do about it all?

(c) I will briefly discuss another global environmental effect: the pollution of water supplies, particularly in the Bengal Basin (West Bengal and Bangladesh) by arsenic and the implications this has for the world.

I conclude with an argument for the future. Contrary to the defeatist (Luddite) attitude toward technology that many non-technologists have, I argue that the future of the human race demands an active intervention by people, to prevent such catastrophes as the Black Death (which probably reduced the population to one third) and all-out nuclear war.

GENERAL INTRODUCTION

I am giving this keynote talk somewhat under false pretenses. I am NOT an expert on climate change. Although, I was for nearly 5 years, Director of the NE Center of the National Institutes for Global Environmental Change (NIGEC) that duty served mostly to tell me how little I know. I have however been aware of the problem for 55 years and

* Mallinckrodt Research Professor of Physics, Harvard University, Cambridge, MA 02138

Global Warming and Energy Policy, Edited by Kursunoglu *et al.*
Kluwer Academic/Plenum Publishers, New York 2001

have written about a number of global issues. I will start by talking about an issue I think I understand - the problems of air pollution, and end with a global problem no one understands - arsenic pollution. Then I will make my appeal: if man wants to live in the world with the population we now have, we must not merely live with the environment we must learn how to manage crucial parts of it - but manage it wisely.

THE ISSUES OF LOW DOSE LINEARITY

Underlying most environmental issues is whether or not there is linearity of effect with dose at low doses. I argue that it is probably true both for radiation and true for many other polluting agents. There is no doubt that high doses of radiation (500 Rems) have led to acute radiation sickness and death; and doses just less than these (100 Rems) have led to cancer. But there is much more doubt whether the low doses that arise from normal operation of nuclear power plants, and even doses to most of the people exposed in accident conditions, lead to any health problems. Since 1928 it has been conventional for prudent public policy, to assume that there is a linear relationship between radiation dose and response so that even small doses, if widely spread over a population, can produce an appreciable response. This assumption was originally suggested (implicitly) by Crowther (1924) and for many years was only made by those concerned about radiation exposure. This led to an (incorrect) feeling that anything involving radiation is uniquely dangerous. It is now realized that this low dose linearity assumption is probably equally valid (or invalid) for other carcinogens, and even other medical outcomes. This is an inherent consequence of the multistage theory of carcinogenesis, particularly in the form developed by Doll and Armitage (1954, 1957).

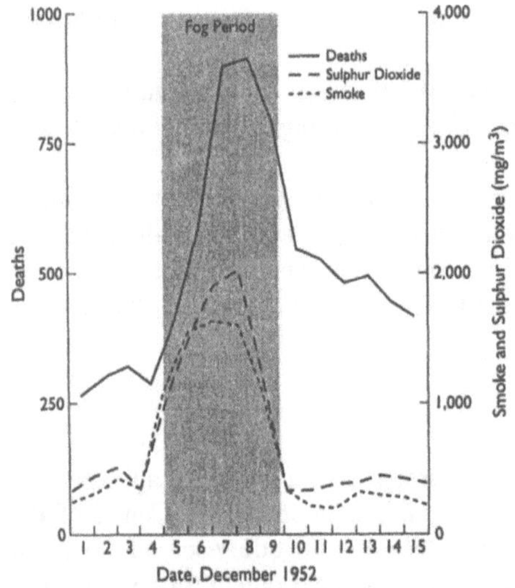

Figure 1. Daily mean pollution concentrations and daily numbers of deaths during the London Fog Episode of 1952. (Source: Beaver, 1954.)

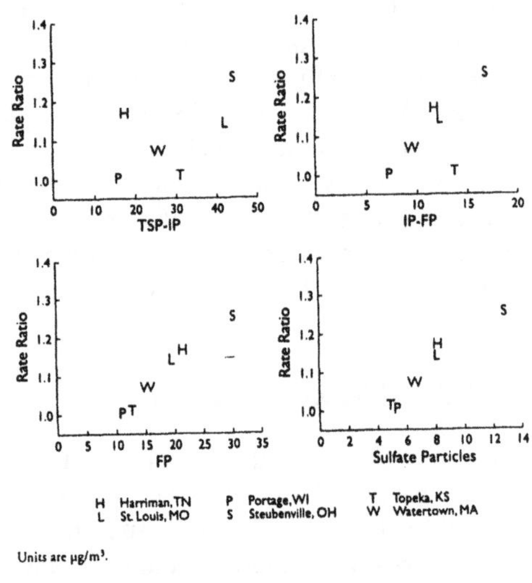

Figure 2. Estimated adjusted mortality rate-ratios from the Six-Cities Study plotted against non-inhalable particles (TSP-IP), the coarse fraction of inhalable particles (IP-FP), fine particles (PM_{25}), and sulfate particles. (Source: Pope and Dockerty, 1996.)

Indeed the idea is more general: if the medical outcome is indistinguishable from one caused by natural processes and the agent acts as the same way as the natural processes at any stage in the carcinogenic process, then almost any biological dose-response relationship will be differentially linear (Crump et al. 1976). Recently it has been realized that linearity may apply to many other situations such as particulate air pollution also (Crawford and Wilson 1976). This is crucial as we reevaluate data on air pollution. In 1925, for example, the cry was "the solution to pollution is dilution", thereby bring all concentrations below an assumed threshold. This certainly reduced local pollution, but increased pollution at a distance and made a local problem into a global problem - albeit one of smaller *individual* concern.

THE EFFECTS OF AIR POLLUTION

The effects of fossil fuel use on public health are primarily those of air pollution: the liberation of gasses from the power plant as a result of fossil fuel burning. Burning of fossil fuels results in emission of gases from incomplete combustion as well as gases from impurities. There is a marked difference between fossil fuel and nuclear plants in these respects. The emissions from fossil fuel plants occur in ordinary operation and are continuous, whereas the only important emissions from nuclear plants occur in accident situations. The pollution from was noticed in the 15th century in England. In the 17th century Evelyn wrote a tract on the subject (Evelyn 1661). There is no doubt that air pollution and in particular the burning of coal HAS killed members of the public outside the power plant when air pollution levels were high. After a large fraction of people got sick and died in bad fogs in Meuse Valley, Belgium, and in Donora, Pennsylvania, people paid attention. Immediately after a London fog in December 1953 there were 4500 "excess deaths" (Beaver 1954). (Figure 1) Deaths were due to a variety of causes all of which can also occur naturally. The British government took the immediate action of banning the burning of soft coal in the cities. The plentiful supply of oil from the Middle East enabled the UK economy to do without this burning.

At that time the UK government's scientific advisors, believing in a threshold, argued that (i) if pollutant levels could be reduced 5 fold the effect would vanish, (ii) the pollution only affected the aged and sick who would die in a few days or weeks anyway and (iii) the major problem was sulphur dioxide from burning of the sulphur pollution in the coal. It is my contention that all three scientific statements (i) (ii) and (iii) were wrong. There has been, and still is, a major controversy about (ii) in particular, and how to extrapolate these known hazards to the lower levels of today. During the 1970s I, and many others, thought that *present day* air pollution in the eastern US or northern Europe affects 1% of the people exposed (Wilson et al. 1982). But in an influential 1979 review (Holland et al. 1979) several prominent British scientists systematically discounted the evidence presented in the studies carried out so far, which merely compared average death rates with averaged outdoor pollution and are classified by epidemiologists as "ecological" studies . They concluded that the health effects of particulate pollution at low concentrations could not be "disentangled" from health effects of other factors. A major (unstated) reason for dismissal of the findings was the (correct) observation that by 1980 visible air pollution in many major cities (London, Glasgow, Pittsburgh, Moscow) had already been much reduced over the black periods of the first half of the century. No government took any further action.

However there is now much stronger scientific justification. Exposure of animals to combustion products begins to elucidate mechanisms (Amdur 1989). Systematic correlation studies of death rates with air pollution in major cities show consistent results. Moreover two epidemiological cohort studies avoid many of the criticisms that applied to the ecological studies. The first was the "Six Cities study" (Dockery et al. 1992) involving a 14-16 year prospective follow-up of 8111 adults living in 6 U.S. cities: Harriman, Tennessee, St. Louis, Missouri, Steubenville, Ohio, Portage, Wisconsin, and Topeka, selected to be representative of the range of particulate air pollution. Measurements were made of total suspended particulates (TSP), PM_{10}, $PM_{2.5}$, SO_4, H^+, SO_2, NO_2, and O_3 levels. Although TSP concentrations dropped over the study periods, fine particulate and sulfate pollution concentrations were relatively constant. The most polluted city was Steubenville; the least polluted cities were Topeka and Portage. Differences in the probability of survival among the cities were statistically significant (P= 0.001). Individual health outcomes were compared with average exterior concentrations. Mortality risks were most strongly associated with cigarette smoking, but after controlling for individual differences in age, sex, cigarette smoking, body mass index, education, and occupational exposure, differences in relative mortality risks across the six cities were strongly associated with air pollution levels in those cities.

These associations, shown in figure 2, are stronger for respirable particles and sulfates, as measured by PM_{10}, $PM_{2.5}$, and SO_4, than for TSP, SO_2, acidity (H^+), or ozone. The association between mortality risk and fine particulate air pollution was consistent and nearly linear, with no apparent "no effects" threshold level. The adjusted total mortality-rate ratio for the most polluted of the cities compared with the least polluted was 1.26 [95% confidence interval 1.08-1.47]. Fine particulate pollution was associated with cardiopulmonary mortality and lung cancer mortality but not with the mortality due to other causes analyzed as a group. The results are substantially larger than those in one of the best earlier "ecological" studies. This difference suggests that the cohort study is able to estimate the effects of air pollution with more accuracy. Ecological (population) studies are averaging the observed effects over the affected population, therefore lower rates in these studies compared to cohort studies would be expected. Although the statistical associations were with area outdoor pollution measurements, subsidiary studies show that gases and fine particles easily penetrate indoors (in contrast to heavier and larger particles). Similar results were observed in a second, larger prospective cohort study.(Pope et al 1994) Approximately 500,000 adults drawn from the American Cancer Society (ACS) Cancer Prevention Study II (CPS-II) who lived in 151 different U.S. metropolitan areas were followed prospectively from 1982 through 1989. Individual risk-factor data and 8 year vital status data were collected. Ambient concentrations of sulfates and fine particles, which are relatively consistent indoors and out were used as indices of exposure to combustion source ambient particulate air pollution. Both fine and sulfate particles are used as indices of combustion source particulate pollution, which is considered by many to be a likely agent. Sulfate and fine particulate air pollution were associated with a difference of approximately 15 to 17 percent between total mortality risks in the most polluted cities and those in the least polluted cities.

These and other, data are summarized in Wilson and Spengler (1996) and have been replicated and reviewed by Krewski et al. (2000). A simple application of these results to the continental USA suggests that 70,000 persons die early (have their lives shortened) by air pollution. I assume that 40% of these, or 30,000 deaths arise from the existing coal fired electricity generation (about 200 GW-yr) to get a coefficient of 150 deaths per GW-yr (Clean Air 2000). *This, if true, dwarfs all other health problems of fuel use.*

A general model that might describe the effect is that the air pollution reduces lung function in an irreversible way. Lung function falls with age, and in the presence of the pollution could fall to a dangerous level when all sorts of ailments occur, at an earlier age than otherwise. This is shown diagramatically in figure 3. It is easy to see geometrically, that the calculated "loss of life expectancy" is directly proportional to the assumed lung damage, and death rate is also proportional. Various studies show that average reduction in lung function is related to air pollution variables although there is large individual variation. This is a delayed effect that is not easy to alter after the initial lung damage. But this is similar to the delayed effect of the cancer mortality in a nuclear power plant accident. The magnitude can be summarized by saying that air pollution, mostly from coal burning, but somewhat from oil burning also, causes more effects on public health than would a Chernobyl-size nuclear accident every year. A group called Clean Air (2000) recently calculated the PM2.5 contributions from power plants based on reported emissions in general agreement with this.

But there remains a huge uncertainty that is related to item (iii) above. Although it is likely that air pollution from fossil fuel burning causes adverse effects on health and likely that there is low dose linearity, it is far less sure what aspect of the air pollution is the problem. I believe that fine particles are involved. But exactly what aspect? Amdur (1989) showed that sulfate particulates are worse than sulphur dioxide for guinea pigs. But sulphur dioxide, emitted as a gas from power plants and evading all the particulate traps, converts to sulphates in the power plant plume as demonstrated unequivocally by measurements from a TVA power plant (McMurry and Zhang 1989). Many countries have therefore been active in reducing sulphur dioxide emissions. But it could also be the vandadium or zinc that often attaches to the particles. A crucial question is whether nitrates are as important (or as good a surrogate) as sulfates -because nitrate precursers are emitted from all automobiles and power plants.

How can one find out? An ideal, but very expensive, epidemiological study would involve following more people (perhaps 100,000) in a prospective study, with each person carrying an personal monitor continuously instead of using the external, area, concentrations. But it is unlikely that this uncertainty will be resolved by epidemiology alone. It will need a careful combination of laboratory, animal and epidemiology experiments to elucidate the probable causes. Meanwhile it behooves us to be careful how we burn fossil fuels. We can, if we wish to spend the money, do a lot about particulate air pollution. At a power plant we can burn natural gas. Natural gas burns quite cleanly. It tends to be free of sulphur and of heavy metals like vanadium and zinc. The temperature can be well controlled so that fewer nitrogen oxides are produced. If we wish to burn coal we can first gasify it and remove these undesirable pollutants.

Once we go beyond the simple exhortation, "avoid burning fossil fuels", or trapping all the particles or particle precursers, the recommendations for public health are far less secure. There is, by now, enough sulphur control that the fine particles are mostly nitrates. If indeed they are as bad as the others, we must be extraordinarily careful about automobile and truck exhausts. Motor vehicles could also be controlled. Catalysts already do a great deal, and we can demand that SUVs and commercial vehicles are similarly equipped. Hybrid, gasoline/electric engines enable the fuel to be burned more cleanly and when fuel cells are used the particulates are more easily removed. Electric engines in automobiles would put all the pollution at the power plant. BUT the overall

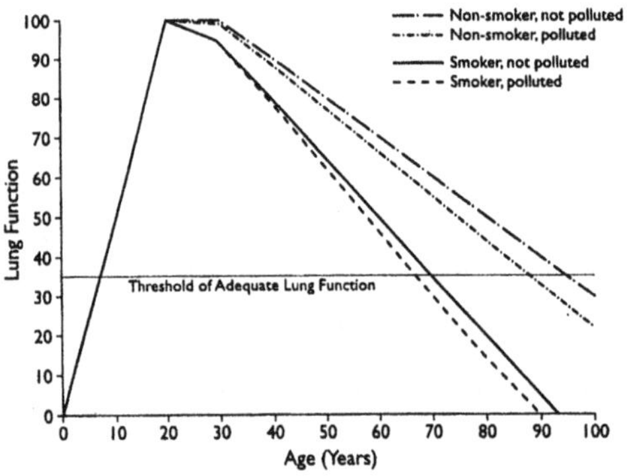

Figure 3. Schematic of lung function vs. age showing loss of life expectancy (LOLE). (Source: Wilson and Spengler, 1996.)

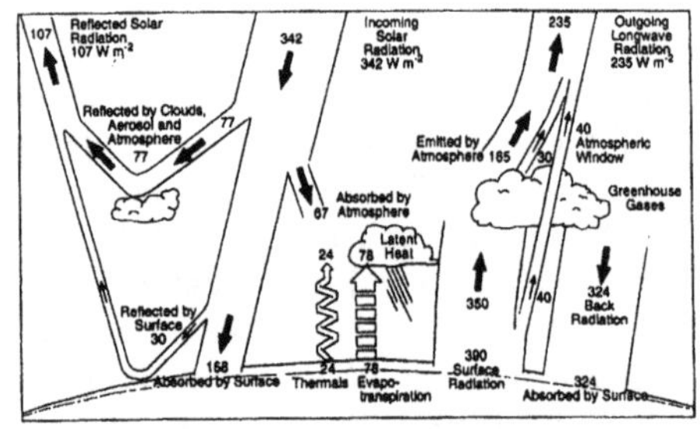

Figure 4. The Earth's annual and global mean energy balance. Of the incoming solar radiation, 49% (168 Wm^{-2}) is absorbed by the surface. That heat is returned to the atmosphere as sensible heat, as evapotranspiration (latent heat) and as thermal infrared radiation. Most of this radiation is absored by the atmosphere, which in turn emits radiation both up and down. The radiation lost to space comes from the cloud tops and atmospheric regions much colder than the surface. This causes a greenhouse effect. (Source: IPCC, 2001.)

simplicity goes down (causing extra expense) and the efficiency may go down, increasing the production of CO_2. And, of course, we can always replace the fossil fuels completely by some other fuel.

HAVE MAN'S ACTIVITIES CAUSED GLOBAL WARMING?

There is a long history of scientific study of global warming. Fourier (1835) may have been the first to notice that the earth is a greenhouse, kept warm by the atmosphere, which reduces the loss of infrared radiation. Without a greenhouse the temperature of the earth would be simply calculated by a balance between the absorption of energy from the sun and black body infrared radiation. It would be about 200 degrees Kelvin, or 245 degrees if some allowance is made for reduced absorption and emissivity. This is too low to sustain life as we know it. With a greenhouse to absorb the infrared radiation from 180 degrees and reemit radiation over 360 degrees, the temperature goes up by a factor equal to the fourth root of 2. Although this picture is still the basis, all sorts of complications set in as illustrated by figure 4 showing the estimates of global energy flows made by the

International Program on Climate Change (IPCC) in 2000.

Even in the 1800s carbon dioxide and water vapor were known to be major "greenhouse gases" since they absorb infrared radiation. They have since been joined by methane, nitrogen oxide, and fluorocarbons. It was also realized that the earth is a leaky greenhouse. Not all the infrared radiation is absorbed by the CO_2. As the CO_2 concentration increases, the greenhouse becomes better and the temperature will rise. Arrhenius (1896) was the first to quantitatively relate the concentration of carbon dioxide (CO_2) in the atmosphere to the global temperature and its changes over the ages. I was first made aware of the problem in the fall of 1947 by the lectures by Dobson to the undergraduate physicists at Oxford. Scientific understanding has increased since then, particularly stimulated in the latter half of this century by the conclusion of Revelle and Suess (1957) that human emissions of CO_2 exceed the rate of uptake by natural sources in the near term and that CO_2 only mixes readily in shallow oceans and mixes with the deep oceans with a 500 year time constant. Then anthropogenic increases in concentration of CO_2 exceed the natural fluctuation. The demonstration by Keeling (1989) (Figure 5a) showed dramatically that atmospheric CO_2 is steadily increasing and the diurnal fluctuations were being exceeded. However, the fluctuations in CO_2 over the millenia were demonstrated clearly by the VOSTOK ice cores taken in the antarctic (Figure 5b).

But the expected effect on temperature was slow to manifest itself. In figure 6, I show the surface temperature record for the last century and a half; although there was an increase of 0.5 degrees in the first half of this century, which occurred *before* the big rise

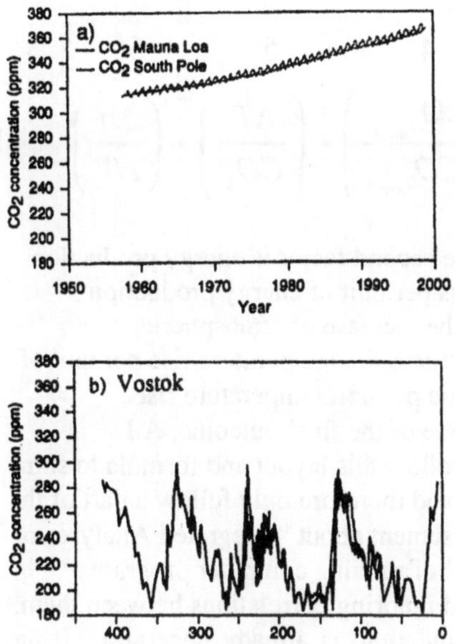

Figure 5. Variations in atmospheric CO_2 concentrations on different time scales. (a) Direct measurements of atmospheric CO_2, and (b) CO_2 concentration in the Vostok Antartic ice core. (Source: IPCC, 2001.).

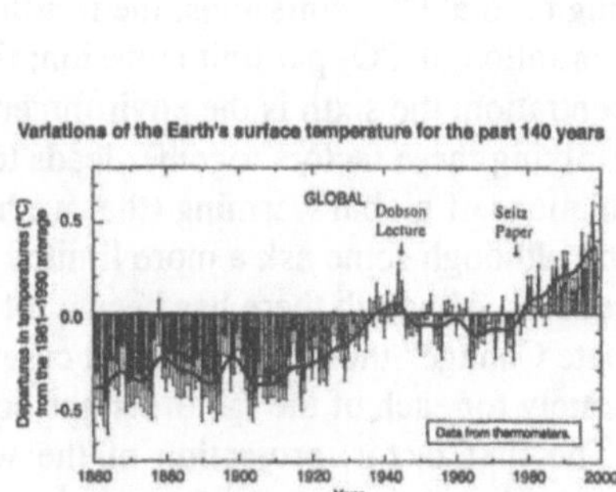

Figure 6. Combined annual land-surface air and sea surface temperature anomalies (°C) 1861 to 1999, relative to 1961 to 1990. Two standard error uncertainties are shown as bars on the annual number. (Source: IPCC, 2001.)

in CO_2 concentrations. I note that when I attended Professor Dobson's lectures in 1947, the rise was still in progress, and there was a slight drop until 1980. That led many people, including some scientists to doubt the scientists' warnings (Seitz 1984). For this and other reasons, the warnings of global warming had little effect on public opinion and policy until the summer of 1988, when it was noted that five out of the previous six summers in the United States were the highest on record and a long-term global temperature record was presented to the U.S. Congress suggesting that a global mean warming had emerged above the background natural variation (Hansen 1981). Although Hansen's specific presentation was based on erroneous data, the more recent rise in temperature shown in figure 6 tends to support his view. There have been criticisms that these data are inconsistent with satellite observations, and therefore unbelievable (Singer 2000). But a special panel of the National Academy of Sciences (NAS 2000) notes that "the warming trend (of the surface) during the last 20 years is undoubtedly real and is substantially greater than the average rate of warming during the 20th century". The warming has been largely at high latitudes and in the center of continents, and is less in the troposphere: facts that were predicted by the global climate computer models.

The core of the *scientific* debate on global warming is about the temperature rise resulting from human activity and in particular from an increase in concentration of greenhouse gases. How well determined are the various parameters? To address this question I go back to a simple equation relating an outcome "h" (which might be the height of the ocean) with the various parameters. This equation has 6 factors. (Shlyakhter et al. 1995)

Factor 1 2 3 4 5 6

$$\Delta h = Popul \bullet \left(\frac{energy}{person}\right) \bullet \left(\frac{CO_{2_{emit}}}{energy}\right) \bullet \left(\frac{CO_{2_{atmos}}}{CO_{2_{emit}}}\right) \bullet \left(\frac{\Delta T}{CO_2}\right) \bullet \left(\frac{\Delta h}{\Delta T}\right) \qquad (1)$$

To begin, the first factor is the world population; the second factor is energy production *per capita*; the third factor is the total CO_2 emissions per unit of energy production leading to total CO_2 emissions; the fourth factor is the increase of atmospheric concentration of CO_2 per unit emission; the fifth factor is the temperature rise per unit of concentration; the sixth is the environmental outcome per unit temperature rise. Multiplying these factors together leads to an estimate of the final outcome. All calculations of global warming (that we have seen follow this layout and formula to some extent, although some ask a more limited question, and therefore only follow a part of the procedure. Although there has been a lot of advertisement about "Integrated Analyses of Climate Change" these have in most cases consisted of running computer programs separately for each of the factors in the equation and ignoring correlations between them.

The first factor, projection of the world's population, is already uncertain. It has been doubling at an increasing rate since 1600. Malthus and others argued that it would go on increasing until war and famine set in. This doubling suggested that there might be a disaster in the year 2000 with a doubling occurring very fast! I am glad to report that this has not happened. Demographers point out that there is a demographic transition. As a society becomes prosperous the birth rate goes down. The net reproduction rate has fallen below 1 in all developed societies and there are signs that it is happening in the people. Stirling Colgate of Los Alamos does not believe this optimistic prediction. He argues that there will always be fringe societies that act differently, and predicts a population explosion about 2070.

The second factor is a major factor in contention. In the USA we use 10 kilowatts per person continuously, in Europe about half this, and in India the amount is 300 watts per person - about a person's internal heat generation rate. We all hope that developing countries will gradually become more prosperous and be able to afford more fuel, but few people anticipate this factor will rise to more that a two or three kilowatts averaged over the world's population by 2100[1]. Even fewer predict that the USA will do what all the rest of the world is encouraging it to do[2] - reduce the energy use per person from 10 kilowatts to 3 kilowatts.

The third, crucial, factor depends almost entirely upon the extent that society will continue to obtain its energy from burning carbon. It is hard to criticize a hope that mankind will know how to interrupt the flow of energy from the sun and put it to good use[3]. But there remain two contrasting views here. According to the first, mankind must use nuclear energy for the foreseeable future if the steady improvement in welfare of the human race is to be maintained. According to the second view that I do not share, we already know enough to use solar energy efficiently and at an affordable price.

The fourth factor depends upon the fate of CO_2 in the environment. Only about half of the CO_2 emitted anthropogenically ends up in the atmosphere. The other half is absorbed in the environment. The factor is about 0.5 and cannot be greater than 1. It is probably known to an accuracy of 30% - better than any other in the equation.

Factor 5 is the main output of General Circulation Models (GCMs), and it is here that the main scientific controversy lies. I argue that it has an uncertainty of a factor of three. We obviously know well the infrared absorption spectra of the various greenhouse gases. If they are uniformly distributed in the atmosphere, and the concentrations are not correlated, then we can "simply" calculate the effect on the greenhouse and hence calculate a temperature rise. Indeed all the greenhouse gases except one, water vapor, distribute almost uniformly around the globe. Although the absorption spectra overlap a little, this is a secondary effect. The IPCC have defined a peculiar concept called "radiative forcing" due to the various gases. This is the effect of each gas separately, assuming an existing temperature and concentration distribution. This can be well estimated for CO_2 , CH_4 and fluorocarbons. Figure 7 shows the relative magnitude of the forcings, and the extent to which they are scientifically understood. If there were no change in the concentration of water vapor (such as would be the case if the Earth was dry), the global-mean surface temperature would increase by $\Delta T_d = 1.2°C$, for a static doubling of CO_2, and this estimate is quite reliable. Simple calculations of the concentrations of these greenhouse gases (except water vapor) from known emissions are, for the most part, well understood. If temperature rise does not affect the concentrations, as is approximately the case for these gases, then the calculation of temperature rise would also be well understood. Also well understood is the fact that the radiative forcing effect of CO_2 is dependent upon concentration. The absorption is complete at the main absorption lines, and as concentration increases it is only at the edge of these lines that absorption increases - leading to a variation as $(\text{concentration})^{0.5}$. This then leads to a reliable estimate that the global-mean surface temperature would increase by $\Delta T_d = 1°C$, for a static doubling CO_2 concentrations. This is an increase in CO_2 concentrations that is maintained at a constant level over a long period of time. This is

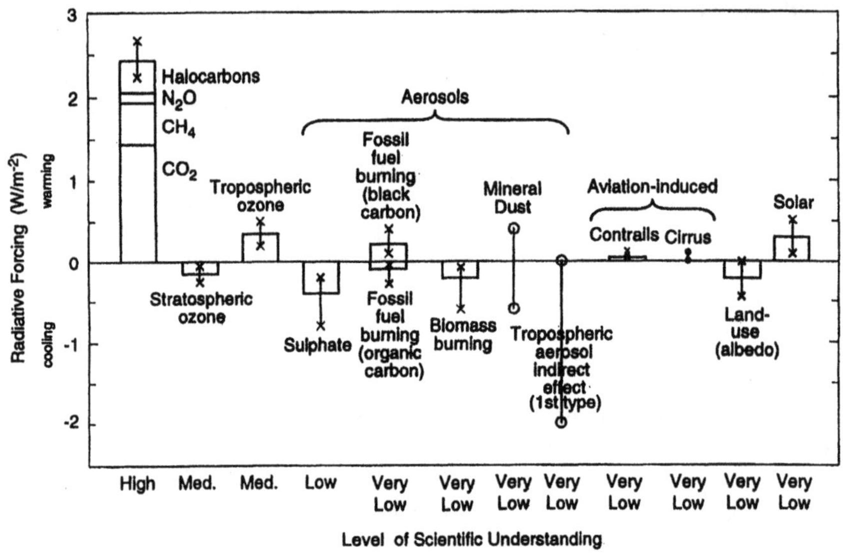

Figure 7. Global, annual-mean radiative forcings (Wm^{-2}) due to a number of agents for the period from pre-industrial (1750) to present (late 1990s: ~ 2000). (Source: IPCC, 2001.)

sometimes called an "equilibrium response" to a static, or quasi-static, doubling of CO_2. an increase in CO_2 concentrations that is maintained at a constant level over a long period of time. This is sometimes called an "equilibrium response" to a static, or quasi-static, doubling of CO_2.

More important, as anyone can see by looking upwards at the world's clouds, the concentrations of the most important greenhouse gas, namely, water vapor, vary rapidly over space and time. If temperature rises slightly because of increased CO_2 concentrations, it is generally assumed that there will be an increase in evaporation of surface water, an increase in water vapor concentrations and hence an amplification of the temperature rise.

Numerous interactive feedbacks from water (most importantly, water vapor, snow-ice albedo, and clouds), introduce considerable uncertainties into the estimates of the mean surface temperature rise, ΔT_s. The value of ΔT_s is roughly related to ΔT_d by the formula $\Delta T_s = \Delta T_d/(1-f)$, where f denotes a sum of all feedbacks. The simple argument above suggests that there is a positive feedback. On the other hand, cloud feedback is the difference between the warming caused by the reduced emission of infrared radiation from the Earth into outer space, and by the cooling through reduced absorption of solar radiation. The net effect is determined by cloud amount, altitude, and cloud water content. As a result, the values of ΔT_s from different models vary from ΔT_s=1.9°C to ΔT_s=5.2°C (Cubasch and Cess 1990). Typical values for these parameters are ΔT_o = 1.2°C and f= 0.7, so that ΔT_d = 4°C. It is important to note that some feedbacks of water vapor may not yet have been identified. In addition the lag of temperature rise is large enough that, with the present rate of increase of CO_2,. at the time that CO_2 doubling is reached, only a 0.5°C - 1.0°C temperature rise is expected, not the equilibrium temperature of 2.5°C. (Cubasch 1992, 1993).

This range is shown in the next two figures. Figure 8 shows a range of predictions in the recent IPCC (2001) report, suggesting that at the end of 2100 there is likely to be a global temperature increase of 5 degrees but it could be as high as 11 degrees. Figure 9 from a more optimistic view by Patrick Michaels and others (Michaels et al. 2000) still shows an increase of over 1 degree by the year 2100. More importantly, all predictions show the rise continuing. Indeed the models all have within them a time delay. In addition, most experts believe that even small increases in the value of *f* could result in a runaway warming not estimated by any of the models, leading, ultimately, to a different stable (or quasi-stable) state of the Earth's climate (Stone, 1993).

Factor 6 is the effect of the temperature rise on the particular societal parameter of interest. This factor is the most uncertain of the lot because it depends upon the most uncertain phenomenon of them all - human behavior. It should be evident that the equation should branch just before factor 6, to allow for different possible outcomes. Alternatively, several equations may be discussed, and the overall outcomes related to each other (perhaps by a cost per unit outcome) and summed.

The crucial issue that faces us in discussion of this factor is man's extraordinary ability to adapt. As global interactions increase, so does the speed of this adaption increase. What is the limit? 20 years ago there were loose predictions of 1 to 10 meter rise in sea level due to global warming as the polar ice caps melt. These have now moderated and predictions are less than 1 meter. But in many locations of the world the local rise, or fall, in sea level due to "natural variations" is more than this. Since The Dutch have shown us, over the centuries, how to cope with a sea level that is higher than the level of the land, should we be worried?

INTERNATIONAL TREATIES

We must continuously remember that attempts to reduce concentrations of carbon dioxide and hence global warming is limited by the fundamental fact that carbon dioxide and other greenhouse gases (other than water vapor) spread widely. My fossil fuel burning affects in some degree every person on the planet. More than any other pollution problem this fact demands that decisions be taken collectively on what action to take. No mechanism proposed so far for such a collective decision of the human race seems satisfactory. On one extreme one could set a limit on the amount of CO_2 per capita that each is permitted to emit, and on another one could set as a base line the amount emitted in recent years with proposed reductions below this amount. Chinese emit much less CO_2 per capita than the world average so tend to prefer the first approach. The USA clearly emits more CO_2 per capita than other countries so tends toward the second, although in the most recent years Congress has not accepted any approach.

The Buenos Aires and Kyoto agreements were attempts to establish a baseline based upon the second approach. President George Bush had a good, and trusted, science advisor in Allan Bromley. Allan made sure that no commitment was made that could not be met. But the commitment to keep emissions of greenhouse gases in the year 2000 to no more than in 1990 was not being met by a restriction on CO_2 but by the already agreed (in Montreal) ban on emissions of fluorocarbons. In Kyoto the US Clinton administration went much further without any clear plan. It is important to realize that in

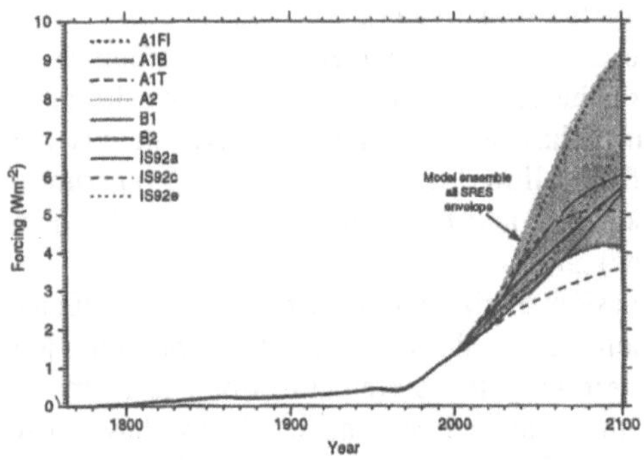

Figure 8.Simple model results: estimated historical anthropogenic radiative forcing up to 2000 followed by radiative forcing for six illustrative SRES scenarios. The shading shows the envelope of forcing that encompasses the full set of 35 SRES scenarios. (Source: IPCC, 2001).

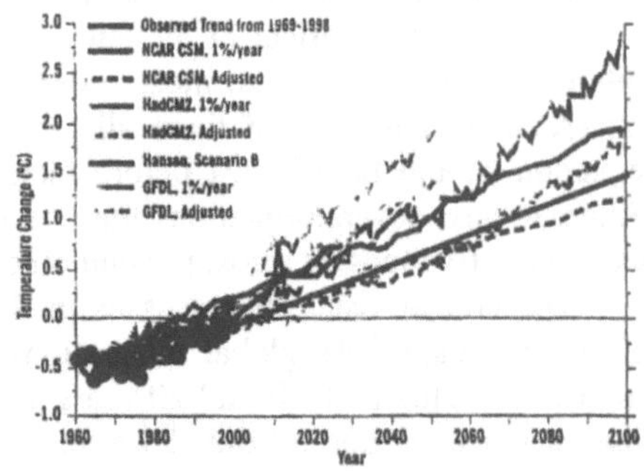

Figure 9. Observed and predicted warming. Note: observed warming of the last three decades, when superimposed on typical climate model projections, shows a linear trend near the lowest value that the climate models predict and considerably below the mean projected warming. (Source: P. Michaels, 2000.)

any ordinary reading of the Kyoto agreement, the USA would have to reduce emissions to 8% or so below 1990 levels by 2008-2012. This is 18% below today's levels. If we allow nuclear power to be abandoned, that is another 5% - a total of 23% which is hard to achieve in ten years. The USA is engaged in unseemly bargaining. We are fortunate in saving a large landmass, which was covered in forests, although some 60% was clear cut by 1860. The regrowth of these forests involves much more uptake of CO_2 Than realized even 15 years ago - expected to account for 8%[4]. But this is an unseemly legal strategy that tends to anger the rest of the world. But we could set in place, if we wish, a national program to expand nuclear power 4-fold. This could not be done by 2012 but a plan could be in place to do so within 20 years so that we would have a good story to tell the rest of the world of a good faith attempt to meet the Kyoto agreement. Vice-President Gore had no such plan. Let us hope that President Bush does.

ECONOMIC INCENTIVES

Once a baseline is accepted, there are still a myriad of options. Let us consider them step by step. We can reduce the world's population (factor 1 of equation 1). We can reduce the energy per capita (factor 2). We can reduce the *net* carbon emissions per capita (factor 3)[5]. We can decide among these, and within these categories, by economic means. Among the economic means suggested for reducing factors 2 or 3 are: penalties for exceeding the emissions quota: emissions trading (both domestic and international), and carbon taxes. All have their problems. There are two moderately distinct ways of altering this factor. One is by improving energy use efficiency, and the other by switching to a non-fossil fuel. Economists argue, with considerable historical justification, that the principal effective way of encouraging energy efficiency is by increasing the price of energy. The use of taxes or charges to reduce energy use per capita has been discussed by Nordhaus and Yohe (1983), Nordhaus (1991), and by Jorgenson and Wilcoxen (1991). Only in one case, mandatory automobile efficiency standards in the USA , compulsion has improved efficiency without a preceding tax, charge, or fuel price increase, although the price of automobiles increased. Another way of raising prices is to allow a cartel to exist. Society, particularly US society, has shown a marked reluctance to do either and politicians have been reluctant to lead.

ACTUAL REDUCTION STEPS

Wilson (1989), Starr (1990), and Bodansky (1990) have all pointed out that for electric power, improvements in end use efficiency (leading to a reduction in factor 2) and a choice of generating source (factor 3) are almost independent societal decisions. The cry often made "we don't need nuclear power if we are more efficient" can be correct only if the desired reduction in fossil fuel burning has a limit. Changing the electric power station to a non-fossil fuel can continue to reduce CO_2 emissions until the last fossil fuel power station is closed.

THE ARSENIC PROBLEM (WILSON 1998).

I conclude by discussing another global environmental problem that I believe has many implications for the future - the problems of chronic arsenic poisoning caused by intensive use of well water. Arsenic has been used since 3000 BC (Partington, 1935). It has long been known to be acutely toxic. Arsenic in water at 60 parts per million (ppm) will kill promptly. The limit on arsenic exposure in drinking water was set primarily to be sure to avoid these acute toxic effects. The arsenic limit set by Bangladesh, the United Kingdom, and the United States is currently set at 50 parts per billion (ppb). Until recently this was also the standard that was recommended by World Health Organization (WHO). But WHO lowered their recommendation to 10 parts per billion (10 ppb) and in the June 22nd 2000 Federal Register the US Environmental Protection Agency formally proposed lowering the standard to 5 parts per billion (5 ppb). This has raised a great furor among those concerned with developing countries. What is it all about?

Arsenic has been used for many years for medicinal purposes. It used to be a cure for diseases such as syphilis and has been shown to assist in curing some leukemias. It was taken as a medicine in "Fowler's Solution" for well over a century. That arsenic at low levels is safe seemed to be reinforced by animal studies that seemed to show that arsenic is beneficial (to animals) at low doses. Indeed the fact that laboratory animals could not be persuaded to develop cancer misled toxicologists throughout the world and was a major contributor to the present catastrophe. But chronic effects of prolonged low level exposure have recently showed up. These include skin pigmentation, and keratoses (and a peculiar disease unique to the area called Blackfoot Disease) which were found by

Tseng in the late 1960s in Taiwan among people who drank from arsenic contaminated wells by Tseng. It was widely thought that there was a threshold at 150 ppb (arsenic in the water) below which no effect was seen. Moreover other organic matter might have been a cause. But Dr Chien-Jen Chen et al (1996) found a very high incidence of lung, bladder and other cancers was found in the same area of Taiwan. In Byrd et al (1996) we plotted, from Chen's data, average cancer rate and average concentration which showed a remarkable straight line through the origin. But we must be cautious. These are "ecological" studies with average rates only. But more definitive "case-control" studies by Ferreccio C, et al. (1998) in Chile and Hopenhavyn-Rich C, et al. (1998a, 1998b) in Cordoba, Argentina, demonstrated unequivocally that there is a 10% risk at levels of arsenic in water only 10 times the prevailing US standard (Figure 10).

Effects have shown up in Inner Mongolia (Luo et al. 1995) where about 300,000 people are exposed above the EPA level and in the Bengal basin where 40 million people are estimated to be overexposed. When I visited Chandripur, a village of 900 people in Bangladesh, 3 years ago, I saw 90 visibly obvious cases of chronic arsenic poisoning. In Wilson (1998) various photographs may be seen of dyspigmentation, keratoses, and skin cancers. The situation is bad enough that I have made an uncontested statement on many occasions, that arsenic in Bangladesh makes Chernobyl seem like a Sunday school picnic. Twenty-five percent of wells drilled in a semi-random manner have high levels of arsenic. The first and most obvious necessity is to measure the arsenic levels in any ground water that is intended for human use. The next step is to purify the water or better still to provide an alternate supply of pure water. The way in which this is done varies from country to country. It is very important to share data and experiences. Although the worst arsenic catastrophe is in Bangladesh, where 35 million people are exposed to levels above the US EPA standard, the amount of arsenic in the soil is less than in many other areas, including areas such as Massachusetts, USA, where it does not appear in unsafe quantities in ground water. It is therefore a very important question as to why and how the arsenic is dissolved in the water. It is therefore important to study the total water chemistry. The chemistry in the wells in Bangladesh is adequate to describe the solution. But the question remains as to when and why the arsenic was dissolved. A clue is given by the measurement in carefully drilled wells that the arsenic concentration is low at the surface, is greatest at about 30 meters below the surface and goes down again at about 100 meters. If a well goes below a clay layer the concentration is al most zero. No other chemical shows such variations. Was the arsenic slowly being dissolved over a million years? or did it only get into the subsurface waters recently? Was the water chemistry changed by the recent rapid pumping to allow the arsenic to enter the water? Or were the natural and man-made changes in river flow, (by the barrage across the Ganges for example) a cause of the problem? These are discussed in several reports and investigations are under way to determine these and other matters.

Why did we not see chronic arsenic poisoning earlier? Why did the world allow this catastrophe to happen - and indeed encouraged the well drilling in Bengal that was the proximate cause of the catastrophe? Why were there reports even as late as 1995 saying that the well water was fit to drink? An examination of the data shows that we DID see the problem earlier and ignored it. Although arsenic was used medicinally in "Fowler's Solution," prolonged use had led to these chronic skin effects. This was reported as early as 1887 by Hutchinson (1887,1888). Figure 11, from Hutchinson's paper, shows pigmentation and skin cancers on patient's hands. A follow up of a number of English

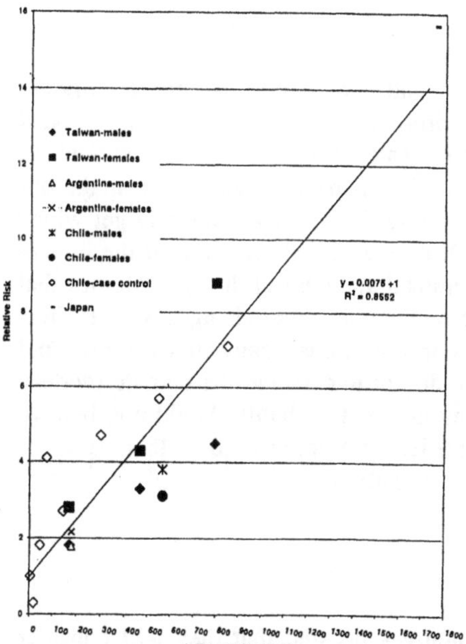

Figure 10. Lung cancer relative risk estimates by arsenic concentration – all studies combined. (Courtesy of A. Smith.)

Figure 11. Pigmentation (left) and skin cancer (right) on patient's hands. (Source: J. Hutchinson, 1888.)

patients treated with Fowler's Solution has been reported by Dr Susan Evans (1983). There were a number of small studies around mines and smelters, which showed effects in a small cohort. They were ignored.

I believe that the main reason for ignoring the problem is that toxicologists and others had used rodents as test animals for 100 years and got used to the idea that if an agent causes cancer in humans it must cause cancer in rodents. Since arsenic does not cause cancer in rodents, ergo, it cannot cause cancer in people! But we now know that it does. The problem is with the toxicologists presumption. This emphasizes the need for constantly questioning the bases for one's thinking. What other presumptions are we making that may later turn out to be incorrect?

The dose response relationship for these internal cancers is consistent with being linear with no threshold. In accordance with my argument at the beginning of the talk, I argue that this should be the default until other detailed information becomes available (Wilson 2000). There are two reasons why the arsenic problem is important. Firstly the world wide exposures to arsenic caused by anthropogenic activity (bringing it to the surface) leads to a known death rate that is observable now, and a theoretically calculated death rate that considerably exceeds that from nuclear power activities and also exceeds that from fossil fuel activities. We must take it seriously. The second is that we must understand the failure to recognize the problem for a century. If we do not, we are likely to repeat the problem with some other, agent, substance or process.

CONCLUSION

In conclusion I note that mankind has now, and always had, a choice. Should mankind allow itself to be ignorant about the environment; to ignore our effect upon it and allow it to control us so that we merely react to whatever happens to us, or should we try to understand the environment, and with great circumspection, learn to manage our local environment in a sustainable way. If we choose the first way, we may get wiped out by the next catastrophe - such as the Black Death that killed one third of the human race. We may get wiped out by a failure to understand and control the technologies that we have unleashed. If we choose the latter we must avoid rapid changes whose effect we do not understand. The recent rise in CO_2 concentrations, many times more rapid than ever seen before, should give us pause. We do not understand the climate models. They do not fully describe past climates. We may not, and probably should not, believe their predictions, and certainly not their detailed predictions. But this lack of understanding should make us more cautious not less cautious.

ENDNOTES

[1] At the Kyoto conference the Chinese representatives insisted that they will consider reducing greenhouse gases when the energy per person is close to the world average - about 3 kilowatts - but not before. It is hard to quarrel with this on moral grounds.

[2] European representatives at the recent (November 2000) Netherlands conference were very insistent on this.

[3] Freeman Dyson, in an optimistic mood, once characterized the various (postulated) societies in the universe. A Class C society can put to good use all the energy arriving from its own sun. A class B society can put to good use all the energy leaving its own sun. A class A society can put to good use all the energy in its own galaxy. We are clearly not yet a class C society.

[4] I regret that the US representatives argued for this at the Netherlands conference.

[5] Although many people believe that it is not economically practical, Dr Klaus Lackner will discuss the possibility of encouraging carbon absorption procedures as well as discouraging carbon emissions.

REFERENCES

Armitage, P., and Doll, R., 1954, The age distribution of cancer and a multistage theory of carcinogenesis, *Brit J. Cancer* **8**:1-12.

Armitage, P. and Doll , R., 1957), A two stage theory of carcinogenesis in relation to the age distribution of cancer, *Brit J. Cancer* **11**:161-169.

Arrhenius, S., 1896, On the influence of carbonic acid in the air upon the temperature of the ground,

Philosophical Magazine **41**:237.

Arsenic Website, 1998) http://phys4.harvard.edu/~wilson/arsenic_project_main.html.

Beaver, (Chairman), 1953, Interim report (on London air pollution incident) *Committee on Air Pollution: Cmd 9011*, Her Majesty's Stationary Office, London.

Byrd, D.M., Roegner, M.L., Griffiths, J.C., Lamm S.H., Grumski, K.S., Wilson, R., Lai, S., 1996, Carcinogenic risks of inorganic arsenic in perspective, *International Arch. Occupational Environmental Health*, **68**:484-494.

Chen, C.J., Chuang, Y.C., You, S.L., and Lin, H.Y., 1986, Retrospective study on malignant neoplasms of bladder, lung and liver in blackfoot disease endemic area in Taiwan" *Br. J. Cancer*, **53**:399-405.

Clean Air, 2000, Death, Disease, And Dirty Power: Mortality And Health Damage Due To Air Pollution From Power Plants, Clean Air Task Force, 77 Summer Street, Boston, MA 02110.

Cohen, B.L., 1977, High level radioactive waste from light-water reactors *Revs. Mod. Phys.* **49**(1):1-21

Crawford, M., and Wilson, R., 1996, Low dose linearity - the rule or the exception?, Human and Ecological Risk Assessment **2**(2): 305-330.

Crowther J., 1924, Some considerations relative to the action of x-rays on tissue cells, *Proc. Roy. Soc. Lond B, Biol. Sci.* **96**:207-211.

Crump, K.S., Hoel, D.G., Langley, C.H., and Peto, R., 1976, Fundamental carcinogenic processes and their implications for low dose risk assessment, *Cancer Res.* **36**:2973-2979.

Dockery, D.W., Pope. C.A., III, Xu, X., Spengler,, J.D., Ware, J.H., Fay, E., Ferris, B.G., and Speizer, F.E., An association between air pollution and mortality in six cities, *New Engl. J. Med.*, **329**:1753.

Evelyn, J., 1661, Fumigorum, or the inconvenience of the aer and smoake of London dissipated. Together with some remedies humbly proposed, 2nd printing, Reprinted by the National Society for Clean Air, London, 1961.

Ferreccio C, et al., 1998, Lung cancer and arsenic exposure in drinking water: a case-control study in northern Chile, *Cad Saude Publica.* **14** (Suppl. 3):193-8.

Fourier, J., 1826.

Holland, W.W., Bennett, A.E., Cameron, I.R., Florey, C. DuV., Leeder, S.R., Schilling, R.S.F, Swan, A.V., and Waller, R.E., 1979, Health effects of particulate pollution: reappraising the evidence, *Amer. J. Epidemiol.* **110**: 525-659.

Hopenhavyn-Rich,C,, et al., 1998, Lung and kidney cancer mortality associated with arsenic in drinking water in Cordoba, Argentina, *Int. J. Epidemiol.* **27**(4):561-9

Hopenhayn-Rich, C., Biggs, M.L., Smith, A.H., Fuchs, A., Bergoglio, R, Tello, E., Nicolli. H., 1996. Bladder cancer mortality associated with arsenic in drinking water and in Cordoba, Argentina, *Epidemiology*, **7**:117-124.

Hutchinson, J., 1887, Arsenic cancer. *Br. Med. J.*, **2**:1280-1281.

Hutchinson, J., 1888, Diseases of the skin: on some examples of arsenic_keratosis of the skin and of arsenic-cancer. *Trans. Pathological Soc. London* **39**:352-363.

IAEA, 2000, Communication from Dr. Abel Gonzales, International Atomic Energy Agency

IPCC, 2000, Draft report from the international program on climate change. Technical summary plus 14 chapters.

Janssen, N.A., Hoek, G., Harssema, H., Brunekreef, B., 1995, A relationship between personal and ambient PM_{10}., *Epidemiology.* **6** (suppl.): S45.

Jorgensen, D. W. and Wilcoxen, P. J., 1991, Reducing US carbon dioxide emissions: the cost of different goals," John F. Kennedy School of Government, Center for Science and International Affairs, Discussion paper number 91-9.

Krewski, D., et al., 2000, Reanalysis of the Harvard Six Cities Study and the American Cancer Society Study of Particulate, Air Pollution and Mortality, A Synopsis of the Particle Epidemiology Reanalysis Project.
HEI Statement;
Preface;
Key portions of the Investigators' Report (Summary of Parts I and II, Introduction, Tables of Contents for Parts I and II);
Commentary by the HEI Health Review Committee;
Comments on the Reanalysis Project from the Original Investigators
Investigators' Report Part I: Replication and Validation
Investigators' Report Part II: Sensitivity Analyses
Health Effects Institute, Cambridge MA, available on the web at
http://www.healtheffects.org/news.htm#Krewski.

Luo, Z-D., Zhang, Y-M., Liang, M.., Zhang, G-Y., Xingzhou, H., Wilson, R., Byrd, D., Griffiths, J., Lai, S., He, L., Grumski, K., Lamm, S.H., 1995, Chronic arsenicism and skin cancer in Inner Mongolia -

consequences of arsenic in well water, *SEGH Presentation Paper,* San Diego June. Printed 5 October.
McMurry, P.H. and Zhang, X.Q., 1989, Size distribution of ambient organic and elemental carbon, *Aerosol Sci.. Technol.*, **10**:430-437.
Malthus, T.R., 1820, *An Essay On The Principle Of Population*, P. Appleman, ed. *Norton*, New York, 1976.
Michaels, P.J., Knappenberger, P.C., and Davis, R.E.., 2000, The way of warming, *Regulation* **23**(3): 10
National Academy of Sciences, 2000, *Reconcling Observations of Global Temperature Change*, National Academy Press. Also available at www.nap.edu/openbook//03090689/html/f.html.
Nordhaus, W.D., 1991, The cost of slowing climate change: a survey, *Energy Journal*, **12**(1):37-64.
Nordhaus, W. D. and Yohe, G. W., 1983, Future Paths of Energy and Carbon Dioxide Emissions, in *Changing Climate. Report of the Carbon Dioxide Assessment Committee*, National Academy of Sciences, Washington, D.C..
Partington, J. R., 1935, *Origins And Development Of Applied Chemistry, Longman*, London.
Pope, C.A., III, Thun, M.J., Namboodiri, M.M., Dockery, D.W., Evans, J.S., et al., 1995, Particulate air pollution as a predictor of mortality in a prospective study of U.S. adults." *Am. J. Respir. Dis. and Crit. Care Med.*, **151**:669-674.
Revellė, R., and Munk, W., 1977, The carbon dioxide cycle and the biosphere, Chapter 10 in: *Energy and Climate, Studies in Geophysics*, National Academy of Sciences, Wash., D.C. pp. 140-158.
Shlyakhter A.I., Valverde, L. J., and Wilson, R., 1995, Integrated risk analysis of global climate change, *Chemosphere* **30**:1585-1618.
Singer, S.F., 2000, Global warming: unfinished business.
Stone, P. H., 1992, Forecast cloudy: the limits of global change models, *Technology Review*, **95**:32-41.
Wilson R., Calome, S., Spengler, J., and Wilson D.G., 1982, *Health Effects of Fossil Fuel Burning*, Ballinger Press, Cambridge, MA.
Wilson, R., and Spengler, J., eds., 1996, *Particles in our Air: Concentrations and Health Effects*, Harvard University Press, Cambridge, MA.
Wilson, R., 1998, The Arsenic Website, http://phys4.harvard.edu/~wilson/arsenic_project_main.html..
Wilson, R., 2000, Comment on proposed regulations (to US Environmental Protection Agency). Available on the web at http://phys4.harvard.edu/~wilson/EPA3_2000.html.

A SCIENTIFIC ASSESSMENT OF EMISSION OF GREENHOUSE GASES INTO THE ATMOSPHERE

FREE-MARKET APPROACHES TO CONTROLLING CARBON DIOXIDE EMISSIONS TO THE ATMOSPHERE

A discussion of the scientific basis

Klaus S. Lackner, Richard Wilson and Hans-Joachim Ziock*

1. Introduction

Human activities are changing the Earth on a global scale affecting virtually every region and every ecosystem [1]. Not all changes have been intentional or for the better. A case in point is the emission of greenhouse gases. There is a growing consensus that the accumulation of greenhouse gases in the atmosphere needs to be curtailed since it has the potential of substantially changing the climate [2]. The dominant greenhouse gas is carbon dioxide (CO_2) generated in the combustion of fossil fuels [3].

Here, we present a trading framework for controlling the introduction of excess carbon into the environment. Trading would limit net carbon influx into the surface pool, which includes the atmosphere, the biosphere and the top layers of the oceans. All these reservoirs are severely strained by the magnitude of the fossil carbon influx. Our goal is to trade introduction of carbon into the surface pool (emission) against carbon removal from the surface pool (sequestration) and gradually achieve a net zero carbon economy. We will discuss the scientific logic on which such trading could be based and the various implications of such a scheme.

At present, there is a lack of political and economic mechanisms to encourage the reduction of greenhouse gas emissions. Consequently emissions keep growing, but with appropriate incentives in place it appears technologically feasible to manage the carbon flux on earth so that carbon would not accumulate in places where it could cause harm.

Preventing the accumulation of CO_2 in the atmosphere is a major challenge to 21st century societies. It requires carbon management strategies that reduce overall carbon consumption, collect carbon from fossil fuel processing power plants, or remove carbon from the atmosphere. In addition to technical solutions, market incentives must be found that provide for cost-effective implementations of these solutions. The realization that fossil energy is readily available and far from running out, together with the observation

* K.S.Lackner and H.-J. Ziock, Los Alamos National Laboratory; R. Wilson, Harvard University.

Global Warming and Energy Policy, Edited by Kursunoglu *et al.*
Kluwer Academic/Plenum Publishers, New York 2001

that fossil energy technology is well developed and cost-effective provides a powerful incentive for solving this problem. [4, 5]

We have developed a scheme of permits, certificates of sequestration, and credits that would be relatively simple to implement and would provide economic incentives for reducing the large net influx of carbon into the atmosphere or reservoirs tightly coupled to the atmosphere. Our goal is to provide the economic mechanisms necessary to achieve a net zero carbon economy.

Controlling Carbon Emissions

Stabilizing atmospheric greenhouse gas levels, as demanded by the 1992 UN Framework Convention on Climate Change from Rio de Janeiro [6], requires a nearly net zero carbon economy. To a good approximation it is the integral and not the rate of emission that causes concern. As the gradient between the atmosphere and adjacent reservoirs decays, holding the CO_2 concentration in the atmosphere constant requires larger and larger emission reductions, which will bring total allowed emissions rapidly down to 30% of what they are today. Ten billion people sharing equally into 30% of the present world CO_2 emission would have a *per capita* emission allowance of roughly 3% of the present *per capita* emission in the United States. Achieving such reductions is beyond the scope of energy conservation and energy efficiency improvements and requires drastic changes in the world's energy sector.

The amount of fossil carbon extracted over the next century is likely to be large compared to the amount that can be readily accepted by the surface reservoirs. These reservoirs, which include the atmosphere, the biosphere and the upper layers of the ocean, need to be considered together as they readily exchange carbon with each other [3]. At current rate of consumption, the emissions of the next century would equal the entire biomass, nearly half of all soil carbon and more than half the amount of CO_2 required to lower the ocean pH everywhere by 0.3.* If the last century is any guide, total carbon consumption could easily be 4 to 5 times larger than the 600 Gt C assumed in this comparison.

Simply prohibiting the use of fossil fuel would not be practical. The world economy depends on it. At present, 85% of all commercial energy is derived from fossil fuels, which are by far the cheapest and most abundant energy resource available. Coal in particular is likely to last for centuries. Estimates indicate a total availability of fossil carbon on the order of 8,000 Gt, roughly 85% in the form of coal [7]. To set the scale, present annual carbon consumption is 6 Gt [8]. Limits on fossil energy derive from environmental concerns not resource availability.

Outlawing emissions to the atmosphere, rather than the use of fossil fuels, might become an option that would allow the use of centralized facilities to generate carbon free energy carriers like electricity and hydrogen. This would require cost-effective means of collecting CO_2 at the source and disposing of it in a permanent and environmentally acceptable way.

---

* Doubling the partial pressure of CO_2 over the ocean surface would in any event cause a decrease in pH by 0.3 in ocean surface water.

Instead of collecting CO_2 at the point of generation, net zero emissions could also be achieved by collecting carbon from any of the surface reservoirs and moving it into long term storage unconnected to the readily interacting surface pools. Such an approach would allow the continued use of carbonaceous fuels for distributed and mobile energy demand. For distributed sources, and in particular vehicles, on board capture of CO_2 is impractical due to the huge infrastructure required for shipping carbon dioxide to central collection points [9]. The only other alternative would be a complete transition to electricity or hydrogen as energy carriers for distributed power applications. A transition to such energy carriers would require a very costly change in the existing infrastructure.

In managing carbon, one needs to consider CO_2 a waste product. In the US the CO_2 output is 22 t/person/year. While there may be limited uses, this amount is far too large for ready use. Avoiding carbon emissions to the air or the disposal of waste CO_2 would not come free. Technologies to approach a zero emission technology in part exist, in part are under development. There are several approaches to collecting CO_2 from concentrated sources like central power plants. Retrofitting seems to incur energy penalties between 10 and 40%, new plant designs could accommodate CO_2 collection without noticeable energy penalties [10]. Integrating CO_2 collection with the removal of other pollutants from the effluent stream may prove very cost-effective [11, 12]. Preventing heavy metals and particulates from a coal-fired power plant from entering the atmosphere may be most readily accomplished in a gasifier plant that also collects all its waste CO_2. Carbon disposal strategies that can cope with large amounts of carbon and keep it out of the surface pool include injection into underground reservoirs [13-16], and the formation of solid mineral carbonates from readily available magnesium silicates [17-19]. The costs of these approaches vary, but they are comparable to the cost of transporting CO_2 to the disposal site, and are likely to end up in the range of a few tens of dollars per ton of CO_2.

This is in stark contrast to some economic models, which consider marginal costs for carbon disposal in the range of hundreds to thousands of dollars per ton of CO_2. New technologies almost certainly will prevent such high prices. Extracting CO_2 back from the air, either through biomass production or dedicated chemical absorption processes [9], also would eliminate the inherent singularity in the marginal cost of complete collection at the source. Collecting all the carbon that is easily captured at a central plant, and making up the difference by collecting in the form of biomass (starting at $25/t of carbon dioxide without credit for energy production [20]) or from the air (roughly $15/t of carbon dioxide [21]) and disposing it by some of the methods mentioned above will lead to a marginal cost without a singularity at net zero emissions. Indeed by extracting CO_2 from the air, it would be possible to operate for some time in a negative net carbon economy. Extraction from the air would also allow the current energy and transportation infrastructure to keep operating through its intended lifetime.

One way of internalizing the cost of managing carbon is through uniformly enforced regulations. Given the nature of the problem this is at best difficult. In any case it would be inefficient, in that optimal sequestration options are not always directly linked to a specific emission. Thus, one needs to find a way of assigning a value to reducing the amount of CO_2 that accumulates in the atmosphere. This could be done either using a tax on carbon consumption or through a permit system where carbon consumption is only allowed after a permit is purchased to cover carbon use. The size of CO_2 emission reductions achieved would be determined by the amount of the tax or the cost of a permit. At first, lower cost solutions would be applied, but these would eventually saturate. At that time, higher cost solutions and further CO_2 emission reductions would only be achieved, if the cost of the tax or permit exceeds the price of further CO_2 emission reductions.

Our approach to internalizing the cost of carbon emissions to the atmosphere is rooted in free-market principles and it has a scientific basis that considers the long-term perspective of the problem. We are concerned with the large amount of carbon that could potentially be moved into the surface pool. In this pool we include the atmosphere, the terrestrial biosphere and the upper layers of the ocean. Each of these has such rapid exchange with the atmosphere that, on a generational time scale, they need to be considered together. Exchange between these pools is largely outside of human control. Even though on average 50% of the carbon emitted to the atmosphere flows into other surface sinks, annual fluctuations can be as large as 100% [22]. As atmospheric carbon levels are stabilized one can expect the outflow into other reservoirs to diminish, as these reservoirs are finite in size and the gradient to maintain a flux gradually decreases.

Rather than attempting to affect short-term atmospheric concentrations, our goal is to develop a platform from which the world can chose the appropriate carbon balance in the surface pool and still retain access to the vast fossil carbon reserves. On long time scales, the differences between various carbon forms become less important. It does not matter whether carbon enters the atmosphere or is instead added to the larger surface pool. Our long-term outlook keeps us from distinguishing between biomass carbon and CO_2 in the air. Also the difference between methane and CO_2 as greenhouse gases becomes a minor complication as methane oxidizes to CO_2 on a short time scale.

The exact impact of accumulation of carbon in the environment is still widely debated. However, as use would grow, the discussion may be made irrelevant by the scale of the emissions. The available fossil carbon dwarfs the capacity of the surface carbon reservoirs that are in contact with the atmosphere. The carbon stored in coal deposits alone is equivalent to about 3,500 ppm of CO_2 in the atmosphere. With increasing levels of CO_2 the debate is likely to shift away from climate change to additional ecological impacts of increased carbon in the surface pool. These will not be limited to effects of atmospheric carbon dioxide but also include the ecological impact of greatly increasing biomass on land and in the ocean as well as the consequences of acidifying the oceans.

The surface carbon pool and sequestered carbon

We distinguish between surface carbon and sequestered carbon. Surface carbon is carbon in the atmosphere or carbon that can readily be exchanged with the atmosphere. Sequestered carbon, in contrast, is well insulated from the atmosphere. Underground reservoirs of carbon are sequestered, so is the carbon in mineral carbonates like limestones and dolomites. Man-made carbonates because of their permanence should also be considered sequestered. Similarly methane hydrates on land or under the ocean are securely sequestered. At what depth in the ocean one can consider organic materials sequestered is likely to be controversial. The distinction ultimately rests on the residence time.

Rather than focusing the debate on greenhouse gas emissions, we submit that in the course of fossil energy consumption sequestered carbon, i.e. fossil fuels, are transformed into surface carbon. On a fifty-year time scale, it is not important whether the carbon enters the surface pool as CO_2, methane or elemental carbon. The ease by which these reservoirs are transformed into each other makes this distinction only of short-term value.

As an example, consider the issue of methane emissions to the atmosphere. Methane is a far more potent greenhouse gas than CO_2 contributing on a molar basis 25 times as much as CO_2 to the greenhouse effect [23]. However, the life time of methane in air is about a decade [22 , 24]. As methane is oxidized to CO_2 it leaves behind little more than the long-term problem of increased carbon dioxide levels. This is typically acknowledged by using a time weighted potency factor, which is about 15 for a fifty year time horizon [22]. If, however, one desires to prevent the greenhouse effect from exceeding some threshold that may be reached in 50 years from now, then the fact that carbon today enters the atmosphere as methane rather than CO_2 is immaterial. If methane alone were the cause of global climate change, curtailing methane emissions would solve the problem in a decade or two. Methane levels are related to methane emissions on a decadal scale not to the integral of the emission. Unless methane emissions grow dramatically, their relative importance as compared to the integral of emitted CO_2 will further diminish. Methane's short lifetime makes a qualitative difference.

One may consider specific reduction targets for long-lived potent greenhouse gases, but one should keep this effort separate from the management of carbon, which is a major player in the long-term greenhouse gas balance, but also introduces other environmental issues that need to be controlled.

The size of the interchangeable surface carbon pool is 42,000 Gt C if one considers the ocean part of this pool. The flexibility or absorption capacity of this pool is far smaller. Allowing doubling in each reservoir but the ocean where we chose an arbitrary limit in acidity, i.e. doubling the H^+ concentration, we are limited to about 3,300 Gt. We consider doubling of environmentally important parameters neither desirable, nor without environmental effect. Yet to hold change below this threshold may be difficult. Rational limits that adequately protect the environment are likely far smaller. Yet 3,300 Gt C is not out of the realm of possible emissions for the coming century.

Our approach to managing the surface carbon budget justifies a permitting approach that is far upstream in the carbon production process and thus is particularly simple. The primary act that requires a permit is the introduction of carbon into the surface carbon pool. Other approaches also consider this point for matter of convenience and practicality, but in our way of looking at it, the most expedient is also the most rational.

The Carbon Board

A governmental body, the Carbon Board, would be given the task of issuing carbon permits. A permit is a document that allows the introduction of a specified mass of carbon into the surface carbon pool. A permit is used once and becomes invalid the moment the associated carbon has entered the surface pool. The Carbon Board should have some discretion in issuing permits and may consider several goals in its permit allocations. The first target would naturally be limiting current CO_2 *emissions*. The second target may be aimed at a desired overall CO_2 *concentration* or *rate of change in concentration*. The desired rate of change could well be negative.* The Carbon Board must consider the supply of carbon permits, and the price of a carbon permit. The Carbon Board needs to assure that permits are always available and it should discourage hoarding. For this reason the board must be careful to avoid issuing an excess of carbon permits. In principle there is no need for an unused permit to expire, but managing the supply may be easier and speculative hoarding discouraged if permits expire.

If the carbon board would adjust supplies so as to keep the price of a permit constant, the permit would effectively act as a carbon tax. If on the other hand the Carbon Board would strictly set a permit supply target, permits would act more like a commodity.

The Carbon Board, like the Federal Reserve Board, would need a certain amount of independence, but at the same time it would need to be guided by general government policy. The Carbon Board would completely control the national supply of permits. On the other hand, demand for permits would change with the state of the economy, changes in energy consuming technologies, changes in the cost of alternative energy, and ultimately in the cost of CO_2 removal from the surface carbon pool. The latter would result in a carbon removal certificate. Such a certificate would replace a carbon permit and thus would limit the demand for permits.

Who needs a permit?

By charging a permit fee at the introduction of sequestered carbon into the surface pool, one minimizes the cost of the regulatory process. Rather than dealing with a myriad of points of CO_2 emission, this approach allows one to charge the fee at an early point in the process chain, where the large scales and simple accounting makes the problem much more tractable than at later stages where there are many emitters and therefore many necessary permits.

Carbon permits are required for mining and other means of extracting fossil fuels. Without international permits, oil and all other fossil fuel imports would require permits when they enter the country.† The oxidation state of the carbon does not matter. Extraction of CO_2 from the ground would require the same carbon permit per mole of carbon as coal. The same is true for CO_2 that leaks from a CO_2 disposal site. The incidental release of methane from a coalmine would need a permit, as would the venting of CO_2 that is stripped from natural gas. Certainly, the flaring of methane at remote sites would require a permit. All the large and easy to measure carbon fluxes should be incorporated by this scheme. Small nuisance emissions could be ignored, as they would raise accounting costs without correspondingly large benefits. One should also note that in issuing permits, one can allow for a multiplier relative to the actual transfers, which

* To achieve a negative rate of change would require certificates of sequestration in addition to permits. These certificates will be discussed below.

† A potentially contentious issue may be the treatment of indirect carbon imports through finished goods.

would allow one to effectively take into account these small nuisance emissions without permitting them separately.

Exempt from permitting are natural transitions from the sequestered reservoir into the surface pool, *e.g.* in volcanic exhalations, natural oil seeps, *etc.* Presumably these natural events are already in equilibrium with counteracting transitions, like natural rock weathering, which, of course, does not generate a certificate of sequestration.

Once introduced into the surface carbon pool, the downstream fate of the carbon does not require any tracking or further permitting. The use of carbon that originated in the surface pool does not require a permit at any stage in the process. For example, the burning of biomass would not require a permit. Combustion of gasoline, regardless of its origin does not require a permit. Extraction required a permit and once the fossil fuel has entered the economy, it, or any of its derivatives, are part of the surface pool and thus do not need to be re-permitted. Losses of carbon, for example from natural gas pipelines, do not require additional permits, even though pipeline leakage may be subject to other regulations.

A permitting scheme based on these simple principles directly interacts with the economy at a small number of points. The cost of the permit would of course be felt throughout the economy. Nearly all activities that require permitting are already subject to careful accounting and thus the permit accounting would not introduce major burdens into the economy. For example the amount of coal that enters the economy is well known and can already be audited. Today, only incidental releases are not yet subject to careful accounting.

Certificates of Carbon Removal

A certificate of carbon removal would be issued for removing carbon from the surface pool and returning it to the sequestered carbon pool. In contrast to the carbon permit, which would always be required when carbon enters the surface pool, a certificate would only be issued if it is applied for, if the transfer can be proven, and if it follows a certified method. A certificate of carbon removal can replace a permit when carbon is introduced into the surface pool. Since certificates of carbon removal are issued for "permanent" disposal, there is no need to limit the lifetime of a certificate. However, it loses its value once it has been applied against a specific carbon emission. The Carbon Board would be responsible for certifying methods of carbon removal as an acceptable means of acquiring certificates. For a method to be certified, it must demonstrably remove carbon from the surface pool and put it back into the sequestered pool. It must do more than what would occur naturally and it must be possible to quantify the additional carbon removed. At the very least, the method must provide a verifiable and reliable lower limit on the carbon removed. Certificates can only be obtained for demonstrated long-term carbon removal. The cost of issuing a certificate of sequestration would have to be born by the applicant.

Instead of moving carbon into the naturally sequestered carbon pool, carbon could be moved into a certified temporary carbon pool. This approach recognizes that not all carbon pools are equally volatile and that carbon remaining in a well-defined location for all practical purposes may be considered sequestered. For a location to be declared as a certified sequestration site, it must be shown that its net carbon flux can be accurately monitored. Once an area is certified and thus subject to carbon management, a net influx of carbon leads to the issuance of carbon removal certificates, whereas a net carbon outflux would necessitate the purchase of additional permits. Accounting may occur, for example, on an annual basis. Given that the cost of permits would be likely to rise with time as the earliest and cheapest CO_2 disposal methods would be adopted first, the responsibility concerning future leakage would strongly discourage questionable short-term solutions. However, financial instruments could be designed to make use of manifestly temporary storage. Rather than introducing a different form of certificate, a temporary method could lead to the immediate issuance of a certificate, followed by a scheduled purchase of permits that over the life of the storage cancels out the initial certificate. The entire transaction could be packaged into a single instrument that not only deals upfront with the re-emission of carbon, but also with the uncertainties of future prices of permits and certificates of carbon removal. Such instruments, although novel, would certainly fall within the scope of present day financial instruments for hedging future risks.

An important consideration for a certified element of the sequestration pool is that its initial baseline carbon content needs to be clearly established. Equally important is that its natural stable carbon content could be reasonably estimated, if at some future date one were to move this reservoir back into the unaccounted surface pool. There needs to be an imputed carbon level for areas leaving the sequestration pool. For example, if one attempts to make grassland a certified element of the sequestration pool by actively farming it to increase the carbon stored as rootmass, credit for artificially high carbon levels in the ground must be returned if the area should be taken out of the sequestered pool. Once the land is not actively maintained, the expectation is that it will revert back to a lower carbon content. The Carbon Board in developing guidelines must assure that the system does not encourage undesirable behaviors like depleting a field of carbon prior to declaring it a certified carbon pool. Simiarly the Board must guard against a ratcheting effect when a reservoir is repeatedly declared a certified carbon pool when it is low in carbon and taken out again when it is high in carbon.

How are carbon permits issued?

There are a number of options to bring permits into circulation, each with its own distinct set of advantages and disadvantages. One we do ***not*** favor would follow the example set by sulfur dioxide emission trading. In that scheme, allowances were assigned to current polluters who could then sell off excess allowances. This resulted in a playing field that was very skewed in favor of entrenched interests. Instead we propose that every party, large or small, new or old, must purchase permits for all carbon it transfers into the surface pool.

The details of the implementation are important. Our approach, although we consider it fairer than the scheme used for sulfur permits, raises the overall cost of carbon, whereas an approach analogous to the sulfur dioxide implementation attempts to limit the cost impact to the margin, where application of technology can make a difference. However, once certificates instead of permits dominate the market, the incremental CO_2 mitigation cost would become the same for both approaches. The advantage of one approach over the other is the time and cost involved in reaching the final state and the perception of fairness, which may be vital to the acceptability of the scheme. Ideally, the money collected for permits would be spent in reducing the cost of CO_2 mitigation efforts. The proceeds could be used to develop vigorous research and development efforts into carbon sequestration, verification and quantification of carbon sequestration and in the development of monitoring systems.

However, it is hard to prescribe such behavior. As a result the approach resembles a carbon tax. It is very important to strive for an implementation that is revenue neutral. Economic modeling by Jorgensen and collaborators and by Manne and Richels suggests that a simple carbon tax (or similarly a permitting scheme) would be deflating to the economy, whereas if it were made revenue neutral by a rebate of another tax it would be stimulating. This suggests a comparable tax reduction, preferably in an energy related field.

Figure 1 makes the points shown above. Curve *A* represents the marginal cost of avoiding a unit of carbon transfer. At the current transfer rate, the unit cost is zero, reflecting the fact that there is currently no financial gain associated with avoiding the introduction of carbon into the surface pool. A permit scheme is then invoked to fix the net amount of carbon that can be introduced into the surface pool. The amount is given by line *C*, the price of a permit is given by line *D*. If one fixes the number of permits, the price of a permit would adjust itself to match the marginal cost of a unit of carbon mitigation. This occurs at the intersection of lines *D* and *C* and curve *A*. *D* represents the marginal unit cost of carbon mitigation. Similarly if one were to hold the price of the permit constant, the number of permits would need to be adjusted so that *C* meets *D* on the curve *A*. At that point, one has reduced the net carbon transfer to the value given by line *C*. Note that the abscissa indicates net carbon introduction into the surface pool and thus does not include carbon that has already been compensated for by some form of sequestration resulting in certificates of sequestration. The integral under the curve *A* between lines *B* and *C*, and denoted by the dark area, is the actual cost to achieve the given reduction of net carbon transfer. The reduction could be achieved by various means including certified sequestration, reduction in consumption, increased energy efficiency and increased energy conservation. Market forces would optimize between the different approaches. The lightly shaded area represents the total revenue from carbon permits. The cost of mitigation and the cost of permits are not in any fixed relationship. At net zero carbon transfer no permits are issued and all carbon is handled

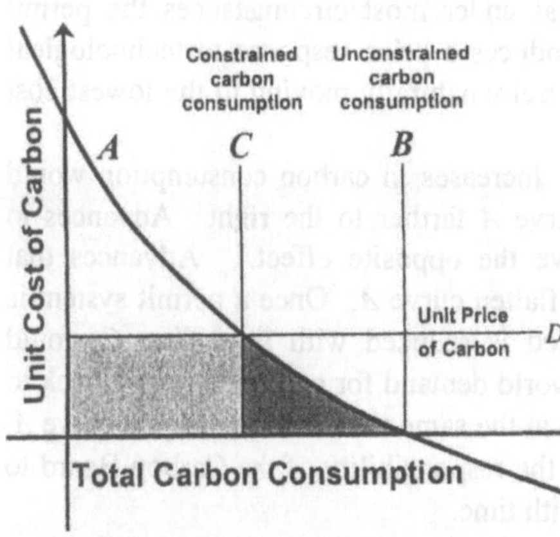

Figure 1: Sketch of unit cost of carbon mitigation, permit prices and amount of carbon reduction. See text for details.

through certificates. However, for small reductions in net global carbon emissions, the revenues collected from the sale of permits would far exceed the money actually needed to reach the level of carbon mitigation achieved.

The way in which the permit system might operate over a period of time can be illustrated by an example. In the first year, permits would be issued for 100% of anticipated demand so that all players would get used to the scheme. In subsequent years, the number of permits would be successively reduced. Further pressure for permit cost increases would result from the trend of continual increases in energy demand. At first the effect would be in promoting efficient use and fuel switching because these are the cheapest technical options. As permits and the work needed to reduce CO_2 emissions become progressively more expensive, sequestration options and alternative forms of energy would begin to be cost effective.

The advantage of bringing the system in slowly is that the market place would have time to adjust. Bringing the system in quickly is likely to be far more expensive. By assigning a large fraction of permits without charge, this cost would be avoided, but it would introduce the questions of who should get these free permits and why.

While a permit scheme controls the supply of net carbon transfers, a tax controls their cost. Rather than controlling the location of the vertical line *C* in Fig. 1, taxes fix the horizontal line *D*. The relationship between fixing the number of permits and levying a carbon tax is akin to that of a free market versus price controls. Based on past experience with price controls, we surmise that under most circumstances the permit system is economically more efficient. It introduces a price response to technological improvements that is lacking in a tax system thereby naturally moving to the lowest cost solution.

Figure 1 would clearly change with time. Increases in carbon consumption would push the crossing of the horizontal axis by curve *A* farther to the right. Advances in technology that increase efficiency would have the opposite effect. Advances that reduce the cost of CO_2 disposal would tend to flatten curve *A*. Once a permit system is put in place and the number of permits issued is reduced with time, line *C* would continually move to the left, unless increasing world demand for carbon pushes it back to the right. The increasing energy demand would at the same time tend to steepen curve *A*. The market would be dynamic and it would be the responsibility of the Carbon Board to ensure that line *C* approaches the vertical axis with time.

The presence of a Carbon Board that can adjust the number of permits to achieve various goals makes the system even more flexible. A Carbon Board that simply fixes the available permits creates a free market with a supply-limited commodity. A Carbon Board that always adjusts the number of permits to hold their price constant has effectively implemented a carbon tax. Other more complex strategies provide intermediate solutions. One possible and appealing approach would be to have the Carbon Board take into consideration the gradual reduction of energy cost that can be expected in a deregulated market.* The price of permits could be adjusted so as to keep overall energy prices constant balancing out reductions in energy prices resulting from

* Notwitstanding the teething pains of the California energy market deregulation, experience elsewhere points to ultimate cost improvements due to increased efficiencies in the market.

increased efficiencies and improved technologies. This would minimize the impact on the overall economy.

Moving to an international approach

The most straightforward international implementation of the permit/certificate system would be to elevate the Carbon Board to the international level and apply the same approach. Two major obstacles render this approach unlikely even though it has some appeal.

The obstacles are the vast disparity of economic strength among countries and the likely resulting redistribution of funds between sovereign countries. At a cost of $10/t of carbon, the permit system would today collect 60 billion dollars worldwide. An International Carbon Board would have to allocate this wealth fairly among all member countries. While the US consumes about 25% of all fossil carbon, it only has 5% of the world population. One might consider different formulas for initially distributing the revenues from permits, but in the end it appears that a fair system would calculate the shares based on a *per capita* basis. However, before *per capita* income is essentially uniform worldwide, this approach would not affect everyone equally and could therefore be viewed as unfair. Alternative schemes might use a distribution formula based on land area, GDP, or even fossil fuel production. Regardless of which approach is finally adopted, real and/or perceived inequities would remain and some groups are likely to object, as they would feel treated unfairly.

Per capita distribution would be most fair in the sense that the benefit derived from energy consumption ultimately affects or should affect the well being of individuals. A similar statement can be made about the detrimental impacts resulting from fossil fuel use. As it is the individual who ultimately pays the cost of the permit, the revenue generated should be allocated on a *per capita* basis. Arguments in favor of other approaches could, however, be made. For example, one can argue that large land areas intrinsically require more energy, especially in the transportation sector, and thus countries in that category may deserve a small break. Similarly, countries with a high GDP act as locomotives for the rest and thus could get a certain break. However, these arguments appear contrived, if they are used for more than fine-tuning a formula that is essentially based on a population count.

One means of implementing a per capita revenue distribution scheme is to print certificates with a country name on each certificate. The fraction of certificates with a given country's name would be based solely on that country's population. The country in question would pay the International Carbon Board for the permits it received, but would then be free to use those permits internally or sell them on the open market. As a result, countries with a small per capita GDP would be assured of a sufficient number of permits to meet their internal needs and be protected against market pressures, which tend to drive the price of permits up on the open market. At the same time they would be able to generate revenue from the sale of their unused permits. Finally at a later date, they would receive back from the International Carbon Board the revenues they paid earlier to receive their permits. Early on this approach would transfer money from the wealthier nations to the poorer ones, providing them with a means for economic improvement.*

A scheme that allots permit revenues on a per capita basis implies a large transfer of wealth from the developed to the developing world. In order to obtain any reduction in CO_2 emissions, the total number of permits worldwide would have to be less than those needed to satisfy current consumption demands. Countries with a small per capita GDP would end up with far more permits than they would need. These countries could in turn sell their excess permits, while countries like the US would need to purchase the vast majority of the permits they need on the international market.

To follow this line of reasoning a little further before abandoning it as impractical, we note that even though the cost of energy would rise in the developing nations, revenues from permit sales would more than compensate for this effect. On the other hand, some of the most vociferous countries like China may find themselves near the boundary where the overall effect is neutral, but as their economy grows, they would become net importers of permits. Energy savings and carbon mitigation would be implemented worldwide, in the developed countries it would reduce the need to import permits, in the developing world it would free more permits for export. The transfer of wealth could be viewed as international aid to countries who need it the most, as measured by per capita GDP or energy consumption which are closely related.

Even though this approach may have some appeal it is not very likely to be implemented, as the economic impact on the industrial nations is very large. It does however raise some interesting issues concerning the developed countries' stand in the Kyoto meeting.

Without an International Carbon Board, an international system could gradually evolve from a collection of national systems. A permit system allows countries one by one to enter into a carbon-controlled economy. One of the major concerns with carbon control of any form is that it puts the country that introduces it at an economic disadvantage. The strength of this approach is that the National Carbon Board can control the cost of a permit and thus keep the disadvantage small. Thus, it allows a country to make the first step without the risk of being left in the cold, if nobody else participates. For a relatively small price, carbon management can be implemented and the National Carbon Board can await international participation before seriously reducing the number of available permits. This we consider a major advantage of our approach.

International negotiations could lead to agreements that oblige National Carbon Boards to set certain policies. For example, the Kyoto treaty could have been couched in these terms.

Permits from a National Carbon Board are by necessity limited to the country of origin. Otherwise countries could issue excess permits to siphon money off the international market. However, an international committee that certifies and supervises sequestration on an international basis could issue international certificates of

* Some precautions must however be taken to avoid abuses. As in the case of national permits, measures must be put in place to prevent hoarding. In addition one must prevent gross inequities in internal redistribution of permit receipts. As an extreme case, consider a ruling class in a poor country directly benefiting by keeping revenues resulting from the sale of permits on the international market while withholding permits and thereby energy from the internal market. One solution may be for the International Carbon Board to sell all permits and allocate the revenues to each participating country based on its population.

sequestration that could be traded in all participating countries. This is possible since certificates are backed by physical removal of carbon from the surface pool. Certificates could even be issued for carbon sequestration undertaken in countries that do not yet use the permit scheme. The availability of international certificates would lead to the convergence of permit prices by adding downward pressure on permit prices in countries with high cost permits. By transferring certificates across international boarders, the efficiency of the process would increase.

Two countries could join into sharing their permit system by merging their carbon boards. This approach is not limited to countries with similar economies, but could work as well between countries of different economic strength. It would, however, imply a financial transfer from the stronger economy to the weaker economy. This may work in cases where an association between countries already exists and other mechanisms are currently used to transfer wealth. A country like Canada could "adopt" a partner country with low GDP. The combined per capita carbon consumption could be close to the world average and the revenue from permits could be assigned in a bilateral agreement.

Unless a joint implementation of this form occurs, it is clear that there is not much incentive for countries with a low per capita GDP to introduce permit schemes unless they feel that they want to keep their per capita carbon emissions much lower than the high GDP countries would do. Thus, the scheme as outlined does not provide any economic incentive for poor countries to reduce their carbon emissions even if it could be easily achieved. In order to provide such an incentive, we introduce the concept of a credit.

Credits

A credit differs from a sequestration certificate. It does not represent removal of carbon from the surface pool, but acknowledges the avoidance of carbon introduction in countries not yet using permits. There is no need to introduce credits in participating countries as permits adequately reward for carbon emission avoidance. As an example of an action that generates credits, consider the implementation of a power plant that is more carbon efficient than some internationally agreed standard, *e.g.* 0.25 kg of carbon per kWh. Every year, the plant would be issued credits for the power output multiplied by the excess carbon efficiency.

On the international market, credits would be used like sequestration certificates. Credit could be given for efficient fossil power plants or the introduction of non-fossil power plants. Other examples include energy efficient infrastructure or the introduction of fuel-efficient vehicles. Not every action that avoids carbon needs to be rewarded by credits, but the international community can decide whether or not it wishes to reward certain behaviors that lower world carbon emissions. In addition, countries that do not use permits could participate through international certificates of sequestration. This would allow such countries to both generate revenue from their CO_2 disposal work and to position themselves for the day when their economies are on firm enough ground to participate in the full international permit system. As economic parity with countries that are part of the existing international permit system is achieved, credits would be reduced and political pressure would increase to join the international system. One approach would be to have the international community to take care of carbon emission related to exported goods and charge a carbon sequestration fee on goods imported from non-participating countries.

Conclusions

This paper is a report on work in progress and it is already clear that there are several items that have to be added to it to be complete. We have deliberately attempted to think the problem through *ab initio* and consequently references to early work are limited.

The present state of the art suggests that technologies for sequestration could become competitive with other means of reducing greenhouse gas emissions. Absent sequestration technologies, the desired reduction in carbon emissions is nearly unattainable. Our framework of tradable permits, certificates of sequestration and credits would allow financial markets to develop that encourage the reduction of carbon emissions. Technical approaches that are encouraged by such a scheme range from simple conservation measures, over energy efficiency and renewable energy to carbon sequestration. Carbon could be captured in centralized plants and immediately be disposed of or it could be recovered from other locations in the environment. This includes carbon capture from the atmosphere.

Recovery of carbon from the atmosphere, whether it is done with biomass or chemical means, provides the one option to leave the existing infrastructure intact and, nevertheless, to manage carbon fluxes. This may prove particularly helpful for automobiles where capture on board is virtually impossible. Methods like these could not function without certificates of sequestration.

Unlike sulfur credits, which represent a right to pollute, CO_2 certificates of sequestration provide a means for brokering between potential sinks and sources. The atmosphere's buffering capacity for the process represents years of world output and thus does not limit this process.

By shaping the market, a carbon board could pursue different strategies for managing atmospheric CO_2. It could ...

(a) fix emission levels at some defined value - perhaps zero;
(b) adjust emissions and CO_2 sinks to fix atmospheric CO_2 *concentration;*.
(c) adjust emissions and sinks to a desired rate of change of concentrations, be they positive or negative;
(d) maintain constant energy prices and rely on improvements in technology to drive down CO_2 emissions;
(e) rapidly expand the issuance of permits if global warming does not take place, or is benign, and one does not need CO_2 sequestration and CO_2 emission limits;
(f) limit pressure on CO_2 emissions such that a country remains competitive with others in similar economic conditions.

We conclude with a possible pathway to a net carbon economy worldwide. It could begin with some of the major industrial countries, which are willing to lead, implementing national permit schemes. To avoid large scale economic distortion between countries, each participating nation would start with plenty of permits available thus making their cost negligible. All which is accomplished in this first phase is the introduction of an accounting scheme. An expiration date on permits would discourage hoarding.

The next phase would see international negotiations encouraged by some or all of the lead countries gradually reducing the availability of permits. Anyone interested in progress can offer a small step forward and await the response of others. If others follow, the next step can be taken. In parallel, one would see the introduction of national and international certificates of sequestration and possibly the introduction of international, or bilateral credits. By working on national and international levels, one can introduce international certificates where agreement can be reached and deal with the more difficult implementations on a local level.

Follow-up agreements on the Kyoto agreement could set targets for national carbon boards to accomplish. National and international use of the funds raised could lead to technology improvements and support for countries that experience economic hardship due to the changes in the energy markets.

Hopefully, the certificates would eventually replace permits on a large scale. As one approaches a net zero carbon economy, further technological advances would start to reverse the cost structure and begin to lower the cost of certificates. Hopefully, more and more countries could reach a level of economic activity that would allow them to participate in the permit scheme. In this stage non-participation would become inconvenient to the country that does not participate, because it is in effect exporting carbon pollution which would have to be reduced by others and quite likely would be charged back to the originator.

As the system approaches a net zero carbon economy, instead of issuing permits, governments could purchase unused credits to further reduce atmospheric carbon dioxide levels. By this avenue, the world could adjust the atmospheric carbon dioxide level to whatever is considered optimal.

List of References

1. Vitousek, P.M., et al., *Human Domination of Earth's Ecosystems* 1997. **277**: p. 494-499.
2. *Climate Change 1995: The Science of Climate Change*, ed. J.T. Houghton, et al. 1996, Cambridge, United Kingdom: Cambridge University Press.
3. Falkowski, P., et al., *The Global Carbon Cycle: A Test of Our Knowledge of Earth as a System*. Science, 2000. **290**: p. 291-296.
4. Edmonds, J., J. Dooley, and J. Clarke. *Carbon Management: The Challenge*. in *Carbon Management: Implications for R&D in the Chemical Sciences*. 2001. Washington, DC: National Research Council. Chemical Sciences Roundtable.
5. Yegulalp, T.M., K.S. Lackner, and H.-J. Ziock. *A review of emerging technologies for sustainable use of coal for power generation*. in *Proceedings of the 6th symposium on (SWEMP 2000) Environmental Issues and Management of Waste in Energy and Mineral Production*. 2000. Calgary, Canada: Balkema, Rotterdam.
6. *United Nations Framework Convention on Climate Change*. 1992.

7. *1995 Energy Statistics Yearbook.* 1997, New York: United Nations.
8. *Energy Outlook 1999.* 1998, Energy Information Administration (EIA): Washington DC.
9. Lackner, K.S., H.-J. Ziock, and P. Grimes. *Carbon Dioxide Extraction from Air: Is it an Option?* in *Proceedings of the 24th International Conference on Coal Utilization & Fuel Systems.* 1999. Clearwater, Florida.
10. Herzog, H., E. Drake, and E. Adams, *CO_2 Capture, Reuse, and Storage Technologies for Mitigating Global Climate Change.* 1997, Energy Laboratory, Massachusetts Institute of Technology: Cambridge Massachusetts.
11. Ziock, H.-J. and K.S. Lackner, *Zero Emission Coal. Contribution to the 5th International Conference on Greenhouse Gas Technologies, Cairns, Australia, August 14-18, 2000.* 2000, Los Alamos National Laboratory: Los Alamos, New Mexico.
12. Feibus, H., *The Future for Coal as A Fuel For Electric Power Generation: A Technical Strategy. Presented at the NEDO Clean Coal Seminar in Hokkaido, Japan, on September 18, 1997.* 1997, Department of Energy.
13. Dunsmore, H.E., *A Geological Perspective on Global Warming and The Possibility of Carbon Dioxide Removal as Calcium Carbonate Mineral.* Energy Convers. Mgmgt, 1992. **33**(5-8): p. 565-572.
14. Hitchon, B., ed. *Aquifer Disposal of Carbon Dioxide.* 1996, Geoscience Publishing Ltd.: Sherwood Park, Alberta Canada.
15. Socolow, R. *Fuels Decarbonization and Carbon Sequestration: Report of a Workshop.* in *Fuels Decarbonatization and Carbon Sequestration.* 1997. Washington DC: Princeton University, the Center for environmental studies.
16. Parson, E.A. and D.W. Keith, *Climate Change: Fossil Fuels Without CO_2 Emissions.* Science, 1998. **282**(5391): p. 1053-1054.
17. Lackner, K.S., et al., *Carbon Dioxide Disposal in Mineral Form: Keeping Coal Competitive.* 1997, Los Alamos National Laboratory: Los Alamos, New Mexico. p. 90.
18. Goff, F. and K.S. Lackner, *Carbon Dioxide Sequestering Using Ultramafic Rocks.* Environmental Geoscience, 1998. **5**(3): p. 89-101.
19. Lackner, K.S., et al., *Carbon Dioxide Disposal in Carbonate Minerals.* Energy, 1995. **20**: p. 1153-1170.
20. Ranney, J.W. and J.H. Cushman, *Energy From Biomass*, in *The Energy Source Book*, R. Howes and A. Fainberg, Editors. 1991, American Institute of Physics: New York.
21. Lackner, K.S., P. Grimes, and H.-J. Ziock, *Carbon Dioxide Extraction from Air?* 1999, Los Alamos National Laboratory: Los Alamos, New Mexico.
22. Schimel, D., et al., *Radiative Forcing of Climate Change*, in *Climate Change 1995: The Science of Climate Change*, J.T. Houghton, et al., Editors. 1996, Cambridge University Press: Cambridge, United Kingdom. p. 65-131.
23. Rodhe, H., *A Comparison of the Contribution of Various Greenhouse Gases to the Greenhouse Effect.* Science, 1990. **248**: p. 1217-1219.
24. Dlugokencky, E.J., et al., *Continuing decline in the growth rate of the atmospheric methane burden.* Nature, 1998. **393**: p. 447-450.

THE PRECAUTIONARY PRINCIPLE:
A guide for action

Jean Couture

1. WHAT IS IT ABOUT?

- A popular expression of defiance? - a set of "good management" rules?

1.1. References to the "precautionary principle" are flourishing in France and in other European countries. The man-in-the-street as well as the media use the phrase all the time, while generally abstaining from defining it. And indeed, most of those who hear or mention it believe they know what it means: they take it as a warning against some kind of danger, implying you have to avoid taking risk.

In fact, it is not so obvious.

Take first "principle". Fortunately, dictionaries are in good agreement in English and for the similar word (principe) in French. Also, in both languages, there are many different shades but you can bring out a few common features: *fundamental doctrine from which others are derived, rule of action, moral law...* . In the popular mind, the word calls for a moral obligation of a primitive kind, imperative in itself.

As for "precautionary", it evokes the presence of *risks,* and the necessity to *avoid them,* to be *secure.* Of course, you may "act" to this end, but it may be hazardous and, if you have choices, the obvious one, more often than not, is to "abstain".

In recent months, the "precautionary principle" is the latest buzzword in daily conversation; good journalists do not hesitate to use it without rhyme or reason. For instance, in a comment of the June 2000 agreement concluded in Germany between the Government and the power industry, you could read in *"Le Monde": "*. . the precautionary principle has prevailed over the realism principle...". In fact, that agreement does not abide by the precautionary principle for several reasons, a strong one being that nobody has a realistic idea of how the power needs of the country might be covered if and when the nuclear plants are closed.

Global Warming and Energy Policy, Edited by Kursunoglu *et al.*
Kluwer Academic/Plenum Publishers, New York 2001

1.2. Undoubtedly, a perception of some kind of risk is at the base of the idea of "precaution". The first movement, in most cases, is to abstain from incurring such risks. But it is also evident that there is nothing like a "zero risk". What is more, it is easy to demonstrate that the "zero risk" concept, practically always, is absurd in itself: if a certain decision involves some risks, abstaining from that decision involves other risks, in one way or other. (This remark is not as superfluous as it may seem: a zero risk being sometimes mentioned or implied in legal documents.)

1.3. So, one can see that applying the precautionary principle, far from being a matter of course, requires a reasoned analysis, taking into account not only the *"pros"* and *"cons"* of the decision in balance, but also the consequences of alternate lines of conduct.

You don't have to be a profound philosopher to find that deciding is often a difficult task. In fact, using the "precautionary principle" consists in following a recipe of good management; and, like most recipes, it never goes without a fair amount of personal judgement. Being human, that judgement is never without some margin of imprecision or error. There is no way to avoid responsibility in taking a decision, with or without principles. All the same, good principles are to be considered and the precautionary principle, if correctly applied, may be quite useful.

1.4. Since a basic purpose of the process is to minimise risks, you have first to make an assessment, as precise as possible, of the relevant risks. How to proceed to that effect is the matter of the next chapter.

2. ESTABLISHED RISKS - ASSUMED RISKS - SITUATIONS OF SCIENTIFIC UNCERTAINTY

2.1. It is important to distinguish between two categories of risk:

Established risks are those which have materialised, even if their occurrence is rare or even exceptional. It is a quite different notion from the probability of those occurrences and from the seriousness of their consequences. Railway accidents, for instance, constitute an established risk. The duty of the management is to take any reasonable measure to reduce both their frequency and the seriousness of the ensuing damage.

When dealing with those risks, you are, so to speak, on firm ground: a scrupulous examination of the circumstances of previous mishap(s) provides some clues on how to avoid similar ones. You may estimate the possible costs of the damage as well as those arising from different sets of measures which may be taken. Those risks and their treatment are not covered by the precautionary principle. For the sake of clarity, it must be agreed that they are to be taken care of by "prevention". To avoid their occurrence and reduce their threat, you face a definite type of mishap. Incidentally, it is reasonable to be prepared for the fact that all you can do will probably be to reduce the frequency of such accidents, not to suppress them.

2.2. The precautionary principle applies to a situation of a different nature: in front of a new problem, e.g. when using a new technology, there is a possibility of some failure which has not, as yet, been encountered. It may be, or not be, something which has a chance of happening. The situation is one of scientific uncertainty. For instance, at the start of the railway industry, some competent people were afraid that tunnels could be dangerous for the travellers, who might suffer from, or even be choked by, excessive air pressure.

When science, at a certain time, is not able to give a definite judgement regarding such possibilities, a decision is necessarily subject to a certain degree of doubt. You may make any step to reduce this doubt, notably through experimentation. But, most often, there is no way of avoiding a residual uncertainty as to the possible existence of such supposed risks.

So, deciding whether or not a new technology may be applied must involve an appraisal of such assumed but unproven risks: how realistic are they? Are they even plausible? What might be the probability of their occurrence? What is the nature and the seriousness of the damage they could provoke? There cannot be a definite answer to such questions. The man or body in charge of the decision must rely on his own expertise, or be content with an answer consisting of an "educated guess" by a reliable expert, or team of experts.

2.3. Then, comes another phase of the process: before deciding for or against a certain proposal, two other questions must be answered:

- what would happen if there is no decision?
- are there other possible solutions and what would their costs and benefits be?

In fact, the two problems are not really distinct: we have noted before that abstaining from a decision is equivalent to taking a negative one. Anyhow, this new phase is certainly an arduous one. In many cases, notably when a decision is mainly (if not formally) of a political nature, it is simply overlooked. In such a case, it is incorrect, even dishonest, to refer to the precautionary principle.

3. EXPERTISE VS. DECISION - THE PRIMACY OF POLITICS

3.1. The precautionary principle may be invoked every time a decision has to be made in the absence of scientific certainty, regarding its possible negative effects. Such a definition is very general and the principle, as previously defined, could apply to disputes between private persons. It might be of interest to discuss such matters, but this is clearly not the right place for doing it.

So let us consider only important decisions to be taken by an official person or body, with public interest in play, e.g. health or environment. To throw some light upon the matter, it is necessary to have recourse to experts. For important projects, several teams of experts are consulted, and of course their advice may differ to some extent.

3.2. Important as it is in the deciding process, the role of the experts must be clearly separated from the decision itself, which belongs solely, at the last stage, to the political authority. In the blame voiced by opponents of decisions they consider unwise or dangerous, there is very often a regrettable confusion: "technocrats" are blamed when their role: expertise, was solely to provide the political authority with data or advice regarding the project.

True, interaction between the experts and the responsible politicians is not always a clear-cut. First, it is up to the latter to determine to which experts they have recourse; but in important cases, the choice is more or less imposed and several sets of advice intervene. Second, it is also true that it is often difficult for a layman to fully understand the language of some experts. It may even happen that a politician - that was recently the case in the French parliament - tries to placate opponents by mentioning that he "always followed the experts' advice"!

Reciprocally, there may also be instances when an expert, aware of some political aspects of the pending decision, is inclined to take those elements of the decision into account in his report. It may even happen that an expert may indulge in some twisting of his work, just to please the man or body who commissioned it.

3.3. So, it cannot be denied that the strict separation between expertise and decision may be somewhat blurred. It remains that the responsibility for a decision fully belongs to the political authority. They cannot argue that "they always followed the experts advice": if they did so, it implies that they could have not followed that advice. Politics has the last word.

4. A GROWING DISTRUST - IS THE WORLD MOVING TOO FAST?

4.1. As we have seen in the first part of this paper, the precautionary principle, in its popular meaning, is the expression of a feeling which is both vague and peremptory: "This new technology might be dangerous, so better abstain from using it."

In such a case, this so-called principle satisfies one of the conditions of a proper definition of the precautionary principle: the existence of a possible danger, even if it is of an unidentified nature. But it lacks many other characteristics, notably an effort to evaluate those assumed risks, the search for possible alternative solutions and a thorough analysis of the costs and benefits of the different possibilities.

4.2. A problem with the popular conception lies in the fact that it is, to a large extent, agreed upon by influential people, notably in the media: journalists like to please their readers. Similarly, and more evidently still, politicians want to please their electorate: such was the case in France when our Government decided to dismantle "Superphoenix", against the advice of practically all knowledgeable people - and, for quite a few politicians involved, against their own judgement.

Another reason for the strength of the "popular" principle is that it sounds like common sense. "Take precautions" before acting is a reasonable rule of conduct, although the meaning of the word is there quite different. And, to add some more wisdom, Latin comes in handily: "In dubio, abstine".

4.3. Certainly, other factors contributed to the coming into fashion of the phrase: it sounds like just common sense, leaving opposite opinions to thoughtless people. Also, the distinction between "precaution" and "prevention" is largely ignored; such confusion may be excusable, since, in most cases, established risks come along with assumed ones. It remains that they are two separate aspects of the problem which cannot and must not be handled as one.

4.4. At the risk of escaping from the framework of this meeting, I would like to loiter somewhat on the popular feeling attached to the precautionary principle.

The explanation of this feeling may be, I believe, summed up in one word: ***contest.*** A new kind of democracy is developing in our countries: every citizen feels he has the right to dispute any decision made by a competent authority, notwithstanding the fact that the latter is entitled to do so by democratically correct rules.

This new "right" leads protesters to gather in *de facto* organisations, who voice their opposition in a forceful way and easily find a resounding confirmation in the media.

Again, such a weakening of the regular authorities requires an explanation. It seems that you could find it in the erosion of popular confidence in the *elite,* especially in science, technology and politics.

A few decades ago, Science was revered as the noble way to secure progress for the people; in French, top science workers were called "les Savants": the men who knew. Today, they are "les scientifiques" and they are included in the distrust which applies to sciences, mothers of nuclear weapons and the GMO.

One may remark that this negative appreciation of scientific innovations remains selective: it does not apply, for instance, to surgery . . . or motor-cars. But its concentration on a few specific subjects, if illogical, seems to strengthen its vigour.

As regards politicians, Italy has opened the way to a new generation of judges who no longer hesitate to investigate the doings of the most powerful people, whatsoever their position may be in business, politics, ... or both. Those judges' counterparts in other countries happily followed. Even if one acknowledges that the judges are perfectly right, their work has disclosed instances of wrongdoing at a frequency and at a level which invite generalisation.

4.5. Admittedly, it seems very arduous to fight popular opposition to many new technologies, especially nuclear ones, and the above analysis does not help much. It may be that scientific and technological progress is too fast: beneficial consequences are quickly integrated and their origin happily forgotten. People remain afraid of change, any change, which is felt as heavy with menacing, unknown dangers.

Is there any remedy? Education is probably <u>the</u> answer, but it would be necessary first to convince educators - a task which looks like a very trying one. Nonetheless, one may question whether the nuclear establishment devotes a sufficient level of effort to that end. *Communication* is the main word of the new century. Business people, industrialists and even scientists have recognised its prime role in human affairs. While I do not claim having any certainty, nor formulating any blame, I feel that the subject deserves a good deal of thought.

How could we reconcile our fellow citizens with technical progress?
Make them as proud of their scientists as they are of their football teams?

5. THE PRECAUTIONARY PRINCIPLE APPLIED TO GLOBAL WARMING

5.1. The precautionary principle is often referred to, in connection with global warming. In fact, few people deny that there are possible risks in the "BAU" (Business As Usual) line of conduct and there is, as of now, no scientific certainty regarding the nature, probability and size of those risks. Some optimists even maintain that natural mechanisms, in the long run, will take care of the problem, so that BAU would be the proper answer.

But a majority of scientists is of a different opinion, although they differ on the middle term and long term consequences of the phenomenon. Therefore, we are, so to speak, in an ideal situation for applying the so often proclaimed "precautionary principle". The sad fact is that it is difficult to find anybody who seems ready to correctly implement the aforesaid principle, as defined by the best possible authors (c.f. annexes).

5.2. Let us remember first that the principle is, in essence, a set of rules and recommendations in view of helping a responsible authority reach a decision in a situation of scientific uncertainty. Now, global warming does deserve its name: the whole planet is really concerned. So there is no political authority in charge of deciding: it is left to international conferences, where governments of sovereign States painfully reach - in the best cases - some agreement. Afterwards, it is up to each country to decide and enforce, as it chooses to do, appropriate rules.

5.3. Expertise is provided by the IPCC (Intergovernmental Panel on Climate Change). Hitherto, they concentrated their work on the extent and possible consequences of global warmmg. Their reports fall short of what is required by the "Ten Commandments".

The first one reads: "Any risk must be defined, evaluated and scaled". Admittedly, it is not an easy task, but it requires a scientific approach and must be the result of a rational process, wholly distinct from the political step that comes afterwards, it cannot be the case with the IPCC. While there is no ground for disputing the competence and character of the governments representatives in that assembly, it is clear that politics cannot be left out of their reports.

Principle VI, at first sight, is applied: "Coming out of uncertainty imposes an obligation of research." Certainly, there are in numberless universities and research centres whole armies of scientists working on the different aspects of global warming. But it is doubtful that, for such a difficult, multifaceted and long-term subject, such an approach be enough. Would not it be appropriate that UNESCO or some other international body put into place a major organisation, something like the "Manhattan Project" with of course input from all the relevant external scientific sources?

5.4. The other side of a rigorous process relates to remedies. What measures should be prescribed in order to alleviate the damages which are dreaded? What would be the costs involved in terms of financial and social problems?

Also, of prime importance in a long term phenomenon, is "Commandment V: How to manage a progressive timing of the strategy in view of adapting it to new elements, progressively provided by experience, scientific and technological developments of all kinds?"

Such a method is evidently out of reach: suffice to mention the North-South opposition. One can all the same regret that, between the developed countries at least, a rational study has not been conducted, with a real will to reach a common analysis, if not common decisions.

True, the Kyoto objectives to limit the GHG emissions were finally agreed upon by all participants: but they resulted from political bargaining. There was no cost-benefit analysis.

5.5. Finally, let me call your attention on the following remark. It is commonly admitted that, even supposing the Kyoto targets reached, global warming will not decrease for quite a number of years. If that is, as it seems, the most likely outcome, why is the man in the street led to believe that the Kyoto agreement's enforcement is a proper way to shield the planet from the dire consequences of a BAU policy? Instead, would not it be more realistic to put emphasis on the possible ways to alleviate the effects of global warming?

Long term measures leading to reduce its causes, necessary indeed, are not the only problem to tackle - perhaps not the most urgent one.

6. SUMMARY

It has been more and more fashionable, in recent months, to advocate the "Precautionary Principle" when dealing with many sensitive subjects, especially in the field of new technologies. In most cases, it is used as a negative argument: "In dubio, abstine".

Popular wisdom may be wrong. There is no way to be safe by "abstaining". A "no" decision is a decision and there is no way to avoid responsibility.

In contrast to the usual negative implication given to it, the real Precautionary Principle is a set of rules which may guide a rigorous process of decision-making. It implies a clear definition of the risks involved, in a situation of scientific uncertainty. Alternative decisions are to be considered, their possible risks and their consequences weighed against each other. Also, costs (in the broadest application of the word) are of course an important factor, which must be clearly estimated for each of the different possible decisions.

Experts are always consulted but the final decision is to be taken by the political authority. The possible role of ordinary citizens and their associations is a difficult problem, since there is a growing tendency, notably in the media, to contest or even oppose any decision which is perceived as a source of risk, while the relentless necessity of taking risks is largely ignored or denied.

Not surprisingly, international bodies are often a long way from applying the real precautionary principle. It is so, even in the favorable case of the EU, a reasonably homogeneous group of Nations.

Lastly, some remarks are devoted to what might be a rational approach to Global Warming, if the precautionary principle were correctly applied to this typical case of scientific uncertainty.

ANNEX I: THE KOURILSKY - VINEY REPORT

It was probably in France that the coming into vogue of the Precautionary Principle was the most acute, to the point where our Prime Minister, apparently tired of seeing the phrase put to every kind of meaning, asked two eminent people, in March 1999, to submit to his Government a "serene, thorough reflexion" on the subject.

Philippe KOURILSKY is Professor at the "College de France", probably the highest title you can think of in the academic world; he is also CEO of l'Institut Pasteur. Genevieve VINEY is Professor of civil rights and the author of several books, concerning responsibility, reparation of damages and compensation for accident victims.

Their elaborate report (about 400 pages, incl. annexes) is conveniently divided, Pr. Kourilsky dealing with the scientific and administrative aspects, while Pr. Viney's comments are of course devoted to the legal questions. It is well specified that both authors are in agreement on the content of the whole book.

However interesting the second part may be - it is quite readable for a non-jurist - ' it would be too much of a task to enter into commenting on its content.

As to Pr. Kourilsky's work, its core may be found in "The Ten Commandments" (see Annex II) which he states should be followed. In a very compact way, they say everything regarding the correct process of decision making, <u>in a situation of scientific uncertainty</u>.

It should be noted that the authors cannot avoid ascertaining that such a process "requires an adaptation of the social actors", which is probably one of the reasons why the Report does not seem to have much impact upon the Government's conduct: any government has heaps of problems to deal with and changing the people's minds is a neither urgent, nor perhaps even realistic, one.

Fortunately, the book has been published and any citizen may buy it. One can hope it will have a positive effect, in the long run, on the public opinion.

ANNEX II: THE TEN COMMANDMENTS OF PRECAUTION

Translated from: Philippe KOURILSKY & Genevieve VIINEY: "Le principe de precaution" (excerpt p.56, Editions Odile JACOB. La Documentation Francaise.

I. Any risk must be defined, evaluated and scaled.

II. The risk analysis must compare the different scenarios of action and inaction.

III. Any risk analysis must comprise an economic analysis which must embark upon a cost/benefit study (broadly understood) before taking a decision.

IV. The structures of risk evaluation must be independent but coordinated.

V. Decisions must, as far as possible, allow for reconsideration and solutions which have been reached be reversible and proportionate.

VI. Coming Out from uncertainty imposes an obligation for research.

VII. Decision circuits and safety devices must be not only appropriate but coherent and efficient.

VIII. Decision circuits and safety devices must be reliable.

IX. Evaluations, decisions and their consequences, as well as any contributory device, must be transparent, which necessitates labeling and traceability.

X. The public must be informed as far as possible, and its degree of participation decided by the political authority.

NUCLEAR POWER – MEETING TOMORROW'S ELECTRICAL GENERATION PARADIGM

Dr. C. K. Paulson*

1. INTRODUCTION

New base-load electrical power generation in many countries today is essentially an oxymoron. Additions in the United States are better described as peaking units as opposed to base generation. Larger plants, whether coal or nuclear, are seldom considered. Additionally, compounding the reigning confusion is the claim that natural gas is "the clean energy." It's a strange time!

Many of the large industrial countries are going through the same deregulation process that is driving the U.S. utility industry, causing them the same pain and agony, and nightmares about the long-term viability of their companies. Other countries suffering from these transition pains are Japan, the United Kingdom, and, yes, even China as they start down the deregulation path.

I am not suggesting that the goal of deregulation is a bad one. However, other fundamental influencing factors must be recognized as requiring significant consideration in future fuel source decisions made by generation companies.

If the universal implementation of deregulation is a principle on which the new paradigm is built, then what are the others that will govern the electrical generation industry decisions in the future?

2. DISCUSSION

First, let me propose a series of "givens" already in place concerning future electrical generation:

1. Renewable energy sources, regardless of their actual costs, are here to stay. You can argue whether this fact is good or bad, but such debate is a waste of time. The powers that have staked their reputations on renewable energy will never back down. This is not necessarily a bad result, even though renewables have

* Dr. C. K. Paulson, Director, Advanced Plant Business Development, Westinghouse Electric Company

Global Warming and Energy Policy, Edited by Kursunoglu *et al.*
Kluwer Academic/Plenum Publishers, New York 2001

their own issues such as high cost, visual impact, regional functionality, etc. Also, to date, the large amount of electrical generation needed to offset current greenhouse gas generators and future generation growth cannot be met utilizing renewable energy sources. However, the results of renewable technology development will lead to diversity of the technology base that will find targets of opportunity uniquely suited for the various concepts. Solar power is a good example of the principle.

2. Despite the overwhelming move towards deregulation around the world, pockets of regulated power will continue. I believe this will be especially true in many developing countries for the foreseeable future. Why is this fact relevant? It suggests that many significant power decisions will continue to reside in the hands of governments as opposed to resting only with the utilities. Therefore, national priorities can cause the principles contained in the paradigm list to change.
3. Environmental impacts will continue to grow as an influencing factor on decisions that deal with technologies that produce or reduce greenhouse gases.
4. Public scrutiny will grow as perceived safety issues gain more media coverage. Nuclear waste, fossil-fuel greenhouse gas emissions, and sensory disparities fall into this category.
5. Large economic investments by electric utilities will become more difficult because of the impact on their balance sheets and profit and loss statements.

The "givens", above, provide a framework on which those of us involved in the new nuclear plant design and construction business believe must be addressed before any possibility of "new-build" opportunities can occur in the U.S and elsewhere. The list, although awesome, may be achievable if the U.S. government does not impede nuclear growth and if the utility industry is willing to participate actively and financially.

3. THE PARADIGM PRINCIPLES

This paper suggest five principles to the electrical generation paradigm:

1. Nuclear power must be competitive with *all* other sources used for electrical generation.
2. The health and financial risks of nuclear power must be similar to other viable electrical generation sources.
3. Nuclear plant safety must be equal or better than other viable methods of electricity generation.
4. The performance of the non-nuclear electrical generation plants must meet the same standards as nuclear plants.
5. The risks of new plant construction must be apportioned based on value-added by the participants.

3.1. Nuclear Power Must be Competitive with *All* Other Sources Used for Electrical Generation

A major criticism of nuclear power has been that generation costs are much higher than fossil energy forms. This comment was not without merit in the past, unless you happen to have bought one of the operating plants that were recently sold at bargain basement prices. The bargain came about because some utilities decided they were not interested in being nuclear operators. Table 1 shows typical generation costs for older, high-performing plants that were hampered during the construction process by high costs due to regulatory changes, construction delays, financial shortfalls, etc. Note that the magnitude of the capital costs significantly impacts the total generation costs. Currently operating nuclear plants are performing extremely well as demonstrated by their low fuel and operation and maintenance costs. However, capital cost impact has been harmful to the viability of nuclear power.

Several situations have the potential to change this high capital cost problem:

1. One-stop-licensing is now an approved regulatory process, thereby significantly simplifying and shortening the regulatory review time.
2. New simplified plant designs promise to significantly reduce the cost of new construction. The impact of plant capital costs on generation costs depends on certain assumptions used in the financial analysis such as internal rate of return (IRR), plant availability, etc. Using typical assumptions, Figure 1 shows the capital cost impacts for various IRRs.

Neither the licensing process nor the new plants designs have been implemented. Therefore, their effectiveness of reducing risks has not been demonstrated. Schedule impacts and actual commodity and labor costs influence the final total plant cost. Therefore, it is critical for a utility to step forward and demonstrate the effectiveness of the new designs and the licensing process. However, detailed cost estimates of advanced plant designs have been completed and accepted by utilities as accurate. Using these cost estimates and projecting generation costs using an 8-percent IRR yields the results shown in Table 2.

Table 1. Generation costs – currently operating nuclear units

Parameter	Cost ($/kwh)
Capital Cost ($2500/kw)	0.0415
Fuel Cost	0.0050
Operations/Maintenance	0.0080
Disposal/Decommissioning	0.0020
Total	0.0565

Table 2. Generation costs – advanced passive light water plant

Parameter	Cost ($/kwh)
Capital Cost	0.0158
Fuel Cost	0.0054
Operations/Maintenance	0.0052
Disposal/Decommissioning	0.0020
Total	0.0284

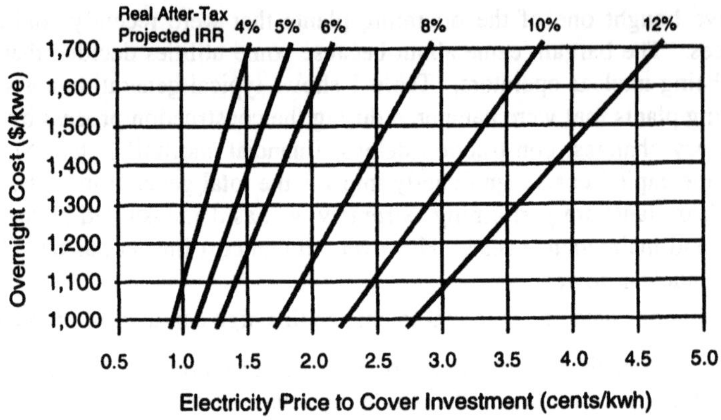

Figure 1. Electricity price required to cover investment costs

3.2. The Health and Financial Risks of Nuclear Power Must be Similar to Other Viable Electrical Generation Sources

Utilities still see the risk of ultimate disposal of fuel as a source of concern – and rightly so. Clearly, adequate technical solutions to nuclear waste disposal exist and are being practiced. Technical risks raised about the long-term storage viability and shipping are exceedingly small when compared to other person-induced risks accepted, and in fact, routinely ignored by the public. (An example of how easily everyday risks associated with power generation are not considered is the expected health impacts of pollution generated by fossil units. The World Health Organization, in their 1997 report on sustainable development, estimates the annual deaths due to indoor and outdoor air pollution from energy production to be approximately 3 million annual global deaths.) However, opposition remains at very high levels of the U.S. government more because of the political support gained rather than any technical concern. The U.S. government Executive Branch position on this subject defies imagination. Until the U.S. government has the fortitude to set a long-term energy policy with clear directives that nuclear power can participate in the future generation mix, final implementation of the waste solution will remain an obstacle to providing a timely and cost-effective solution that offers the potential to significantly offset the environmental damage already attributed to human generated greenhouse gases.

The dichotomy of the position is obvious – reduce greenhouse gases significantly, but don't use the only currently available technology that can economically accomplish this reduction in the foreseeable future.

The potential beneficial impacts of using nuclear power to reduce greenhouse gases are overwhelming. Figure 2 shows a familiar chart depicting the differences in carbon emissions from fossil fuels compared to nuclear. The benefits with respect to greenhouse gas reductions are obvious. However, not so obvious is the magnitude of the "solution space" which, arguably, herein assumes an objective of meeting the spirit of the Kyoto Protocol emission levels. Figure 3 identifies the need to offset about 100 gigawatts of fossil-fueled electrical generating plants within the next 10 to 12 years to approximately meet the Kyoto goal. This also presumes no growth of electrical generation being met by fossil fuels over that time period. Although the U.S. has no chance of meeting this goal,

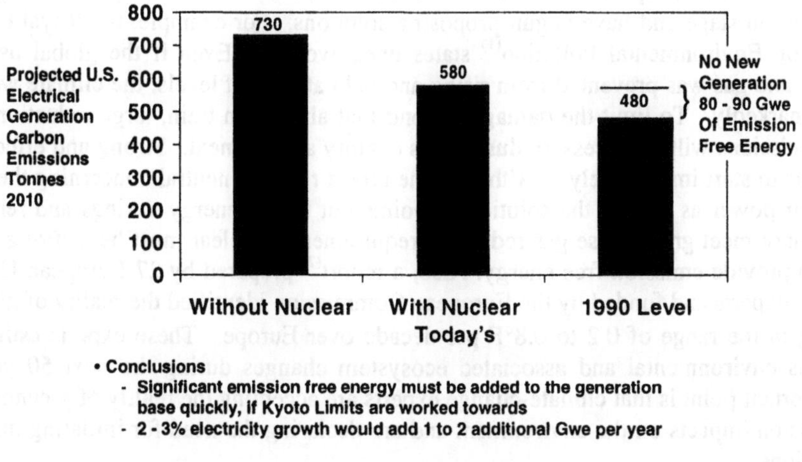

Figure 2. Kyoto emission limits demand significant U.S. actions

any significant progress is unlikely without some commitment to maintain a viable nuclear program.

The ruse of uncertainty concerning human contributions to the global warming appears to be a rationale used to prevent any action on moving ahead with CO_2 limitations. Although our current U.S. president has recently proposed limits on CO_2, the implementation and timing are questionable. Also, the impact on the environment will not be measurable for years. Recognizing global warming as real and committing to a viable solution seems to have at least one similarity to alcoholism or cancer – that is, admitting there is a problem seems to be the necessary step to recovery.

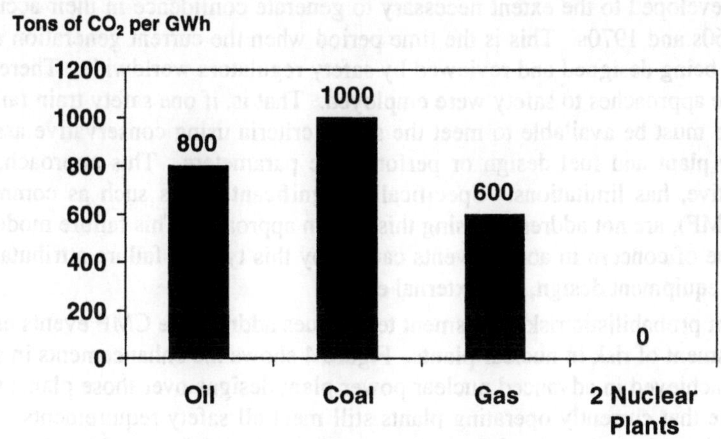

Figure 3. CO_2 emissions comparison

Other countries, especially certain European ones, have been willing to get beyond the admission stage and have begun proposing solutions. For example, the Royal Commission on Environmental Pollution[1] states unequivocally "Even if the global use of coal, oil, and gas was prevented from rising and held at current levels, the climate would change markedly. To limit the damage beyond that already in train, large reductions in global emissions will be necessary during this century and the next. Strong and effective action has to start immediately." Although the report remains neutral concerning the use of nuclear power as part of the solution, it points out that if energy savings and renewables cannot meet greenhouse gas reduction requirements, nuclear must be a strong candidate to provide emission-free energy. Also, a report[2] prepared by 27 European Union scientific experts and funded by the European Commission identified the reality of global warming in the range of 0.2 to 0.8°F per decade over Europe. These experts estimate numerous environmental and associated ecosystem changes during the next 50 years. The important point is that climate-change experts are accepting the reality of greenhouse gas emission impacts on the environment and are declaring the need for initiating immediate actions.

3.3. Nuclear Plant Safety Must be Equal or Better Than Other Viable Methods of Electricity Generation

Recognizing that a proof that this statement will rely on the willingness of the skeptic to be convinced based on qualitative and quantitative data, the following conclusions frame a dilemma:

1. Nuclear designers have made significant progress in improving plant safety.
2. Tools to predict risks of other forms of electrical energy production have not been utilized as extensively as in the nuclear industry.

First, let me mention the significant advancements in measures of nuclear safety that have occurred over the last 15 to 20 years. Risk assessment tools, although not new, were not developed to the extent necessary to generate confidence in their accuracy during the 1960s and 1970s. This is the time period when the current generation of nuclear plants was being designed and reviewed by safety regulators worldwide. Therefore, very prescriptive approaches to safety were employed. That is, if one safety train fails, at least one backup must be available to meet the safety criteria using conservative assumptions concerning plant and fuel design or performance parameters. This approach, although very effective, has limitations. Specifically, significant issues such as common mode failures (CMF), are not addressed using this design approach. This failure mode has been shown to be of concern in actual events caused by this type of failure attributable to human error, equipment design, and external events.

Current probabilistic risk assessment techniques address the CMF events as part of a total assessment of risk in nuclear plants. Figure 4 shows the enhancements in safety that have been achieved in advanced nuclear power plant designs over those plants now operating. Note that currently operating plants still meet all safety requirements. However, advanced plants have benefited from the experience base of operating plants and have been able to simplify and enhance designs to reduce the risk associated with a core damage event by a factor of 10 to 100. The magnitude depends on the design that is being used for comparison. With risks quantitatively defined as once in every 30 million

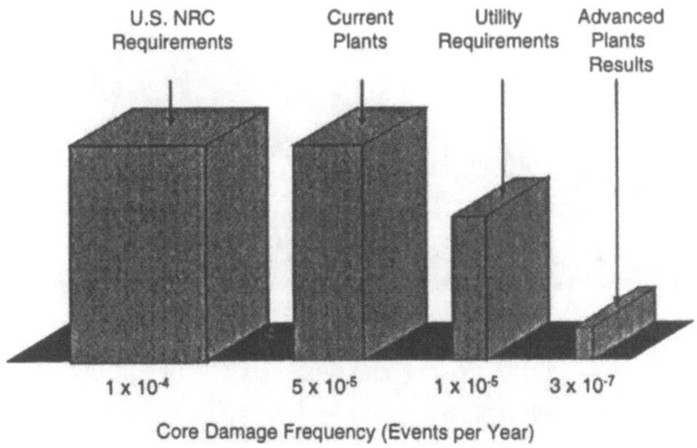

Figure 4. Safety and investment protection

reactor years, the public should feel very comfortable with nuclear plant safety. This is especially true because the risk of a damaged core ever escaping from its vessel is essentially non-existent due to unique design features of some advanced plants.

3.4. The Performance of the Non-Nuclear Electrical Generation Plants Must Meet the Same Standards as Nuclear Plants

The focus in the above statement must be on equivalence. This has not happened in past comparisons. The issue of waste disposal, for example, is treated explicitly when total nuclear generation costs are determined. Also, a cost per kilowatt hour is identified for plant decommissioning and included in the total nuclear generation cost. The adequacy of the amount assumed for these costs components is arguable only when environmental costs are factored into generation costs of competing fuel types.

The impact of greenhouse gas emissions is becoming quantified through implementation of the Clean Air Act in the U.S. by dealing financial credits for reducing NOX and SOX.

As mentioned previously, the U.S president has called for new legislation also limiting CO_2 and mercury emissions. However, inclusion of CO_2 into federal legislation has a long way to go before approval. Credits for CO_2 reductions (or taxes on not meeting reduction targets) have been proposed in other countries. Typical cost adders, per ton of carbon emissions, range from \$20 to \$50, and represent a fairly noticeable impact on the cost of electrical production. Figure 5 demonstrates the potential impact of carbon credits or taxes on generation costs. It also compares the results to the expected performance of the next-generation nuclear plants. Unfortunately, the greenhouse gas reduction from nuclear power plant operations is not allowed as part of the Clean Air Act as an acceptable way to gain financial credits.

This is not true if these reductions occur because of improved performance of the utilities' fossil units. In addition, allowing nuclear credits within the framework of the Kyoto Protocol is being challenged by several European countries. The basis of this

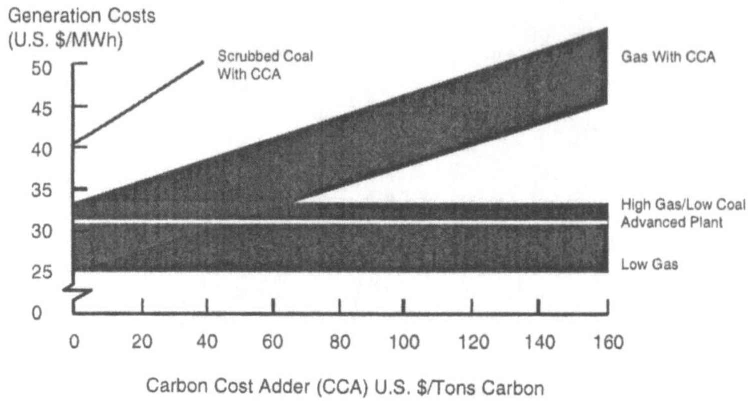

Figure 5. Representative levelized generation costs

antinuclear bias is clearly pressure from the Green Movement, which forces its issues without identifying solutions that can be implemented cost effectively or in a timely fashion.

3.5. The Risks of New Plant Construction Must be Apportioned Based on Value-Added by the Participants

The new paradigm requires a more equitable sharing of the risks associated with re-starting the construction of new nuclear units. In the past, the understanding of which organization was best suited to take on risk was not well thought out, possibly because it was not well defined. The prior generation of nuclear plants was, in many respects, the first generation of nuclear plants. Much has been learned concerning where risks of large construction projects reside. No participant in the project should be put in the position of taking risk that it does not control. However, a participant must be willing to take the risk associated with its own value-added portion. Examples include: engineering companies take design risk, construction companies take risks associated with labor rates, installation rates, productivity, etc.

Utilities should assume those risks associated with plant financing, public acceptance, and site characterization. Predefined contingency funds should be identified prior to start of construction for unexpected costs, and utilized by participants on a direct cost basis with no profit motive.

With the focus on developing a detailed win-win scenario, and shared downside risk, all parties should approach the contract with a positive, "team" attitude.

4. CONCLUSION

The time is ripe for new nuclear construction. Natural gas prices are no longer drifting downward. Advanced nuclear designs have the potential to generate electricity competitively without environmental credits, and may improve substantially if appropriate credit is given for reductions in greenhouse gas emissions. Enhancements in nuclear plant designs have led to a large reduction in risk to the public. In addition, public acceptance suggests a majority of Americans support nuclear power especially when the pollution reduction benefits are understood.[3]

The elements of a successful solution space are within the grasp of the electrical generation industry, but no single company can complete the necessary moves on its own. That's good! The endpoint of re-establishing nuclear power as part of the future electrical generation mix requires the buy-in of all segments of the electrical generation industry.

5. REFERENCES

1. "Energy – The Changing Climate," Royal Commission on Environmental Pollution, Report to Parliament, June 2000.
2. "Assessment of Potential Effects and Adaptation for Climate Change in Europe": The Europe ACACIA Project, Parry, M.L. (Editor).
3. "Perspective on Public Opinion" Prepared by the Nuclear Energy Institute, November 2000

NUCLEAR ENERGY AND ENVIRONMENT: FACTS AND MYTHS

NUCLEAR ENERGY IN THE 21ST CENTURY

Potential for alleviating greenhouse gas emissions and saving fossil fuels

Leonard L. Bennett and C. Pierre Zaleski*

1. INTRODUCTION

Electricity has specific characteristics that make it attractive for many end-use purposes. It is versatile, easy to distribute, clean and efficient at the end-use point and has some non-substitutable uses (e.g., lighting, communications, computers, electric motors). Electrification of rural areas in developing countries contributes to a better distribution of employment opportunities and a more equitable access to health and education services, as well as improving the overall standard of living.

Industrialised countries represent less than 25% of the world population but consume more than 75% of the electricity generated world-wide. On average, electricity consumption per capita is around 500 kWh per year in developing countries, compared to more than 5,000 kWh in Europe (but much higher in Scandinavian countries) and more than 10,000 kWh in Canada and the USA [1].

At present, the electricity demand growth rate world-wide is over 3.2% per year, which is more than twice the population growth rate, slightly higher than the economic growth rate, and almost twice the primary energy demand growth rate. Although electricity consumption in developing countries experienced a 13-fold increase since 1960, consumption is still constrained by supply limitations (one-third of the world population is still deprived of access to electricity), and demand is expected to continue increasing dramatically. Even with strong efforts directed towards demand-side management and reduction of the energy intensity of national economies, the World Energy Council (WEC) estimates that demand for electricity in the developing regions of the world will grow by a factor of 2-2.5 up to 2020 and by a factor of 4.5-6.5 up to 2050, relative to 1990 [2]. In OECD countries, on the other hand, electricity demand is estimated to increase only by a factor of 1.2-2 up to 2020 and to remain relatively flat from 2020 to 2050.

---

* Centre de Géopolitique de l'Énergie et des Matières Premières, University of Paris Dauphine, Paris, France

Global Warming and Energy Policy, Edited by Kursunoglu *et al.*
Kluwer Academic/Plenum Publishers, New York 2001

2. SUSTAINABLE DEVELOPMENT

The concept of *sustainable development* was introduced in the 1980s [3] and gained momentum through the United Nations Conference on Environment and Development (UNCED) held in Rio de Janeiro (Brazil) in 1992. *Agenda 21* [4], adopted by UNCED, called for development strategies aiming towards environmental protection and inter-generation equity, and emphasised that environmental and development concerns should be integrated into the process of decision making. The Second Assessment Report [5] of the Intergovernmental Panel on Climate Change (IPCC) emphasised that mitigation options for alleviating the risks of global climate change should be assessed comprehensively and that adequate policies should be implemented to promote the installation of environmentally benign energy conversion technologies. The Kyoto Conference in December 1997 adopted a Protocol to the Framework Convention on Climate Change (FCCC) aiming towards lowering overall emissions of greenhouse gases. The Protocol [6] will entail industrialised countries reducing their collective emissions of greenhouse gases by 5.2% in the time frame 2008-2012, relative to levels in 1990. However, it has to be noted that ratification of this commitment is lagging, and that implementation of the Protocol is uncertain.

Sustainable development *does not mean* limiting economic growth. On the contrary, a healthy economy is better able to generate the necessary economic resources for environmental protection and improvement. It also does not mean that every aspect of the present environment should be preserved at all cost, but rather that decisions throughout society be taken with proper regard to their environmental impact [7, 8, 9, 10, 11, 12, 13, 14]. Sustainable development *does mean* taking responsibility for policies and actions. Decisions by utility managers, the government or the public must be based on the best possible scientific information and analysis of risk. When the consequences of a decision potentially are serious, precautionary decisions are desirable, even in the face of uncertainty about the consequences. In this regard, however, it should be noted that all alternative courses of action have to be examined, taking into account that a decision on one course of action (e.g., for the phase-out of nuclear energy) will lead to other actions (e.g., to expand fossil fuel burning). Thus, the trade-offs among alternatives must be examined carefully.

Particular care must be taken in cases where the effects may be irreversible. Cost implications should be communicated clearly to the people responsible, and the proper tools of analysis have to be applied in support of policy making.

The concept of sustainable development therefore implies new rules and procedures in the process of electric system planning. Economic, social, health and environmental considerations (some of them long term and/or global) will have to be integrated into the decision making process. The full implications of long term electricity supply strategies have to be considered in an integrated system analysis framework, as opposed to project-by-project decisions. What previously was a process carried out by the electric utility, with the concurrence or tacit approval of the regulatory authorities, now has become an iterative process involving many actors (public, utility, regulators and financial market). The process of enhanced electricity system analysis can be understood therefore as an integral part of a collaborative process that involves all interested and affected parties (IAPs). Such enhanced approaches to electricity system analysis are examined in a reference book published recently by the IAEA [15].

In the power sector, this new concept calls for incorporating three fundamental principles in energy/electricity/environment analysis: economic efficiency; sustainability; precaution.

3. THE PRESENT STUDY

The results and information presented in this paper are drawn from a study [16] carried out by an expert group organised by the International Atomic Energy Agency (IAEA) and the Nuclear Energy Agency of the Organisation for Economic Co-operation and Development (NEA/OECD), with the objective of developing technically sound information for their member countries and for contributing to the work of other international fora, in particular the IPCC and the Subsidiary Bodies to the FCCC. In order to achieve this overall objective, the study encompassed two main tasks:

- development of plausible scenarios of the evolution of global and regional energy and electricity demands in the long-term future (to 2050, with general perspectives out to 2100);
- analysis of the potential role of different energy sources (fossil, nuclear and renewable) and energy conversion technologies in global and regional energy and electricity supply strategies (energy mixes) that are compatible with the goals of long-term sustainable development.

3.1 Study Approach

The approach adopted for the study consisted of forming an international group of experts (see Annex), who worked on an *ad hoc* basis over a period of approximately one year. The study group was organised jointly by the IAEA and the NEA/OECD. Modelling and computational support for the study were provided by experts from the following national institutes: Los Alamos National Laboratory (LANL, USA); Energy Systems Institute of Russian Academy of Sciences (SEI*, Russia); University of Tokyo (UT, Japan).

In addition, advisory support was provided by experts from the Centre de Géopolitique de l'Énergie et des Matières Premières (Centre of Geopolitics of Energy and Raw Materials) at the Université Paris Dauphine (University of Paris–Dauphine, France), and an IAEA consultant.

3.2 Computer Models Used

Each of the national institutes participating in the study used computer models that had been developed, or adopted and modified, by the respective institute, as follows:

- LANL: Edmonds, Riley, Barns Model (ERB), adopted and modified by LANL;
- SEI: Global Energy Model (GEM10R), developed by SEI;
- UT: Linear Dynamic New Earth 21 Model (LDNE21), developed by UT.

The purpose of applying three models to the study is to draw on the complementary features of each. For example, the ERB model uses a 'top-down' econometric approach to derive 'market clearing' energy supply-demand balance (i.e., supply = demand) at each time step, which allows the effect of energy prices on energy demand (i.e., price elasticity of demand) to be taken into account. While the other two models, which use a 'bottom-up' technological approach, do not allow this

* Formerly the Siberian Energy Institute; still referred to as SEI.

effect to be analysed, they complement the ERB model by permitting a greater level of detail in the description of the technical and economic characteristics of different energy conservation and supply technologies. All three models permit the input data to be changed readily, thereby facilitating sensitivity studies. An overall comparison of the key features of the models is presented in Table 1.

In the present study, the final and secondary energy demands were generated by the ERB model, and the results were used as input to the GEM10R and LDNE21 models, which require these data as a starting point for the calculations of primary energy demands and supply mixes (i.e., amount of primary energy demand that is supplied by each energy source and conversion technology). Main inputs to the ERB model, the flow of information from ERB to GEM10R and LDNE21, and principal outputs from the models are shown in Figure 1.

3.3 Description of Scenarios Selected for the Study

The study examined two contrasting scenarios of overall energy demand. The first scenario is referred to as 'business-as-usual' (BAU), and assumes that future energy demand growth will not be governed by policy measures aiming specifically towards protecting the environment. For this scenario, the data on long term trends in GDP per capita, which is an important driving factor for energy demand, were drawn from the IIASA/WEC Scenario B (Ref. 2). The second energy demand scenario, referred to as 'ecologically driven' (ED), takes the contrasting view that specific environmental protection measures will be implemented aiming towards reducing risks of global warming. In this scenario, the GDP per capita data were based on the IIASA/WEC Scenario C.

For each of the two energy demand scenarios, the study considered two contrasting scenarios for nuclear power. The first nuclear scenario, referred to as the 'basic option' (BO), assumes that the growth in nuclear electricity production (non-electrical applications of nuclear energy were not considered) will be driven by the economic competitiveness of nuclear power in comparison with other electricity generation options. In this scenario, the factors influencing nuclear power evolution are consistent with the medium variant (MV) in a recent IAEA/NEA study (Ref. 13). The second scenario for nuclear power, referred to as 'phase-out' (PO), assumes that nuclear power will be essentially phased out of electricity generation by around the middle of the next century, irrespective of its economic competitiveness, driven by national decisions to turn away from nuclear energy. This nuclear scenario is consistent with the low variant (LV) in the IAEA/NEA study (Ref. 13).

The combination of two demand scenarios and two nuclear power scenarios leads to four energy demand-supply scenarios, as illustrated in Figure 2.

It is emphasised that neither the energy demand scenarios nor the nuclear power supply scenarios are intended to be predictive. The only purpose for constructing such scenarios within the study is to examine the effect that such contrasting views of energy demand and nuclear power supply would have on the derived costs of energy supply and on CO_2 emissions from the energy sector.

3.4 Main Drivers of Energy Demand Growth

The main drivers of future energy demand growth are the expected increases in global population, especially in the presently developing countries, and growth of national economies. To the extent feasible, the same data for these drivers were used in all three models applied in this study (for each region treated in the respective model).

Table 1. Overall comparison of the models used in the study

ATTRIBUTE	MODEL		
	ERB	LDNE21	GEM10R
Paradigm	"Top-Down" (Econometric)	"Bottom-Up" (Technological)	"Bottom-Up" (Technological)
Energy resource data - Fossil - Nuclear	 Edmonds/Rogner [17] OECD/NEA [18]	 Rogner [17] OECD/NEA [18]	 Rogner [17] OECD/NEA [18]
Regions (micro/macro)	13/3	10/3	10/3
Time Divisions - Base year - Horizon year - Increments (yrs)	 1990 2095 14	 1990 2100 10	 1990 2100 25
Objective function or convergence criteria	World fossil fuel prices	Total discounted energy cost over time horizon	Total discounted energy cost over each period
Discount rate	5%/yr (when used)	5%/yr	5%/yr
FE demand structure (1) PE and SE demand structure same in all models	Res. /Comm,, Industry, Transportation	NA, SE demand from ERB	NA, SE demand from ERB (2)
Population Data	World Bank [19] analytic fit	NA	World Bank [19] direct input
Per-capita GDP - Source of data - Kind (3)	 IIASA/WEC [2], indirect fit MER	 NA NA	 IIASA/WEC [2], direct input PPP
Technology cost data - Electricity - Other	 NEA [20]/ERB ERB	 NEA [20] LDNE21	 NEA [20] GEM10R
Plant efficiencies - Nuclear - Gas - Coal	 0.35 → 0.40 0.40 0.40	 0.35 0.46 → 0.51 0.39 → 0.50	 0.33 → 0.35 0.45 → 0.55 0.36 → 0.50
Renewable energy (RE) - Commercial - Non-commercial	 Yes No	 Yes No	 Yes Yes
AEEI (4)	1975-2005, 1.00%/yr 2005-2050, 0.50%/yr 2050-2095, linear to 0.8%/yr	NA	NA

(1) FE = Final Energy [energy used by consumers]
PE = Primary Energy [gas, oil, solids (coal+biomass), nuclear, hydro, renewable]
SE = Secondary Energy [gas, liquids, solids, electricity]

(2) Transformed from ERB economic sectors to 4 energy forms: heat, mechanical, electrical, chemical feedstock

(3) Kind: MER = Market Exchange Rate; PPP = Purchasing Power Parity

(4) AEEI = Autonomous Energy Efficiency Improvement. The rate at which the efficiency of converting Secondary Energy (SE) to Final Energy (FE) improves through technological progress, without specific policy measures aiming towards promoting efficiency improvements.

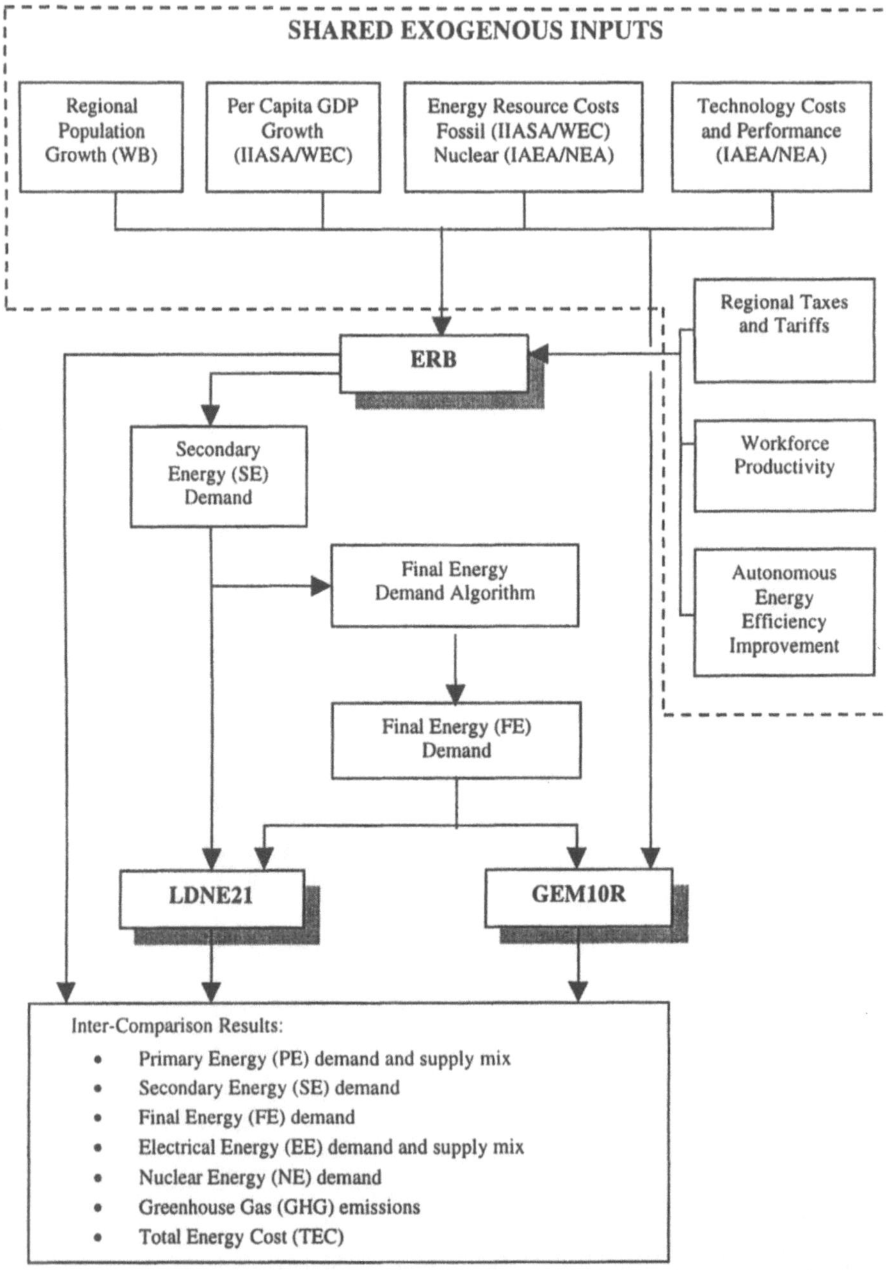

Figure 1. Flow of information in the three-model approach

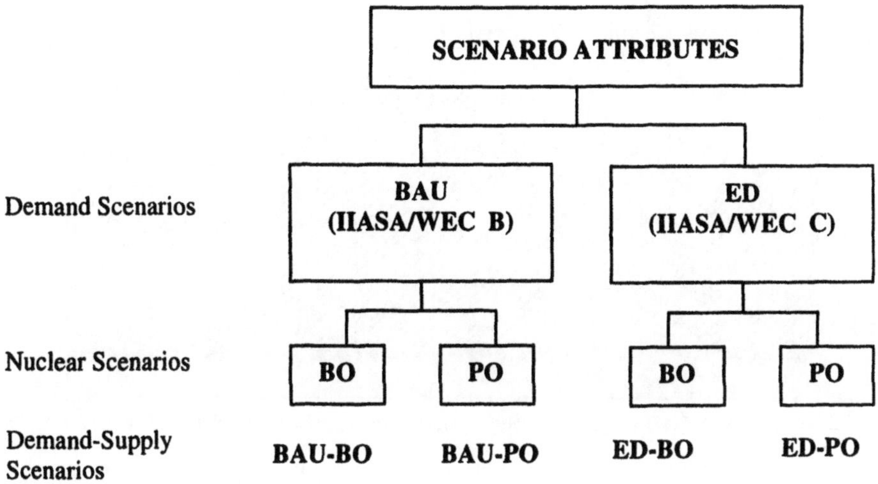

Notes:

Demand Scenarios:	BAU=Business-As-Usual;	ED=Ecologically Driven
Nuclear Scenarios:	BO=Basic Option;	PO=Phase-Out
IIASA/WEC B:	Scenario B in the IIASA/WEC Study	
IIASA/WEC C:	Scenario C in the IIASA/WEC Study	

Figure 2. Energy demand and nuclear power scenarios analysed in the study

The future trends in population, derived from Ref. 19, are shown in Figure 3. As can be seen, world population is expected to grow from some six billion persons at present to around ten billion at the end of the next century. Almost all of the four billion additional inhabitants of the world will be in the countries in the region designated as 'Rest of the World (ROW)'; that is, countries presently at the lower end of the economic spectrum. The population of this region is projected to increase by a factor of about 1.8 by the year 2100. The populations of the OECD group of countries and of the group of countries presently undergoing economic reform (REF) are projected to remain essentially stable through the next century, with the population of the OECD having a small increase and that of the REF region undergoing a slight decrease.

Trends in economic growth (see Figure 4), as reflected in rising gross domestic product (GDP), were derived from the per capita GDP data of Ref. 2. As can be seen, the total GDP (expressed in market exchange rate - MER) of the OECD region is projected to increase by about a factor of three, from some 22 trillion[1] dollars at present to about 66 trillion dollars in the year 2100; that of the REF region is projected to increase by slightly more than a factor of 3, from about 3 trillion dollars today to around 10 trillion dollars at the end of the next century; and the GDP of the ROW group of countries is projected to increase by a factor of about 15, from about 6 trillion dollars today to about 90 trillion dollars in 2100.

(1) One trillion = one thousand billion (10^{12}).

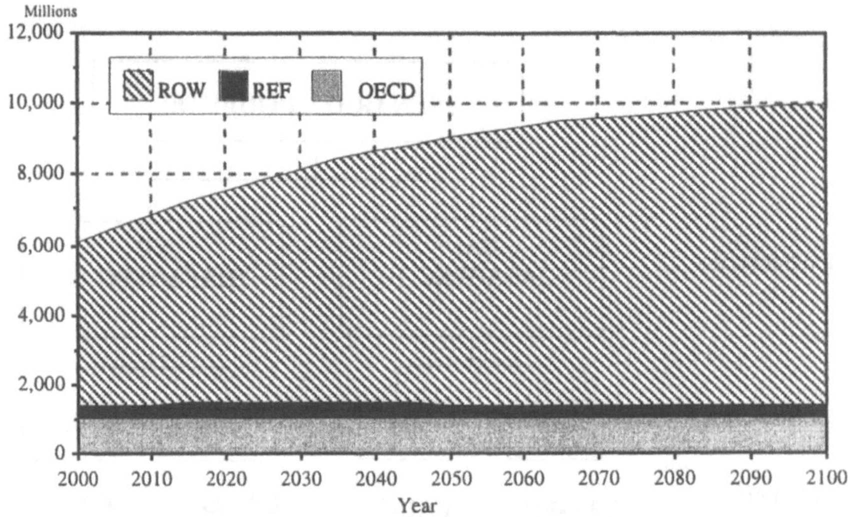

Figure 3. Trends in future population growth (millions of inhabitants)

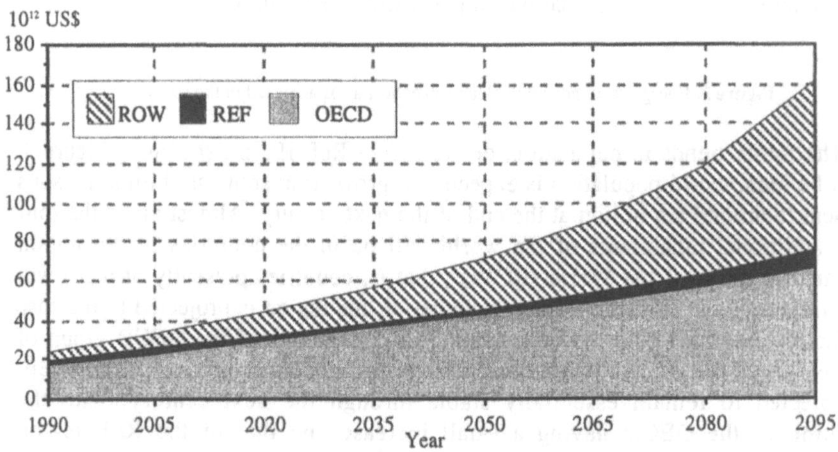

Figure 4. Projections of future global GDP growth (trillion US$ - MER)

However, the 'economic welfare' as measured by GDP per capita shows a rather different trend, owing to the population growth trends presented in Figure 3. As may be seen in Figure 5, the GDP per capita in the OECD region is projected to about triple up to the end of the next century, reaching a level of around US$ 65 000 per person; in the REF region the GDP per capita also grows by a factor of about three but, owing to the present low level of GDP in this region, by the end of the next century the level is projected to be only about US$ 25 000 per person, or about 60% below the level projected for the OECD region. Even though the total GDP of the ROW region is projected to increase by a factor of about 15 up to the end of the next century, the population is estimated to grow by a factor of 1.8 over the same period; thus, the per capita GDP will increase by only a factor of somewhat more than 8, reaching only some US$ 10 000 per capita. Therefore, at the end of the next century, the 'economic welfare' of ROW countries is projected to be still far below (a factor of 7, compared to a factor of 15 today) that of OECD countries.

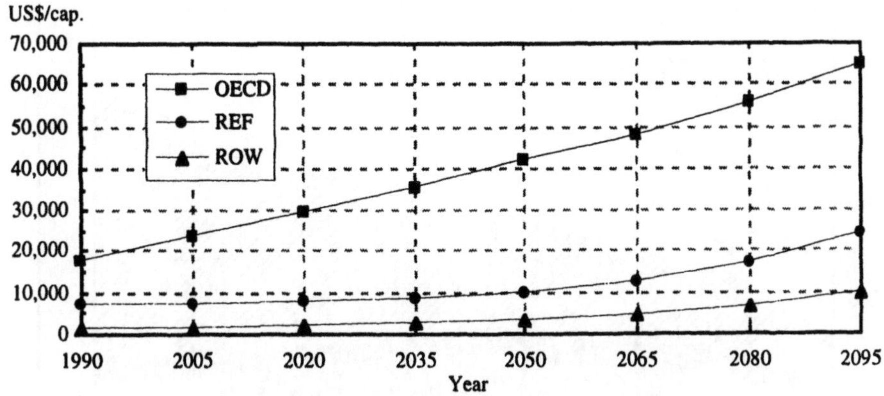

Figure 5. Trends in GDP per capita (US$/capita - MER)

4. RESULTS OBTAINED IN THE STUDY

4.1 Energy Demand Estimates for the Four Scenarios

The starting point in calculating energy demands is the estimation of final (end-use) energy needed to provide the levels of energy services required to meet the rising economic levels of a growing global and regional population. Figures 6a and 6b show the final energy demand results from the ERB model for the four scenarios, at the level of the three macro-regions adopted for reporting results in this study. The following trends can be seen:

OECD: FE demands are projected to increase moderately (some 15%) in the two BAU scenarios up to the middle of the next century, and to decline thereafter to a level some 10% below that in the year 2000. In the two 'ecologically driven' (ED) scenarios, FE demands are projected to remain essentially stable up to the middle of the next century, and to decline thereafter to a level in the year 2100 that is 25%-30% below that in the year 2000.

REF: FE demands remain relatively stable up to the year 2100 in all scenarios, undergoing a small (some 10%) decrease up to the middle of the next century and an increase thereafter, with the demand in the year 2100 being about the same as (the two BAU scenarios), or slightly lower than (the two ED scenarios), the demand in the year 2000.

ROW: in this region, which encompasses the presently developing countries and which is the region of continuing population growth, FE demands are projected to increase by a factor of 3 to 4 by the year 2100. Therefore, the challenge in the next century will be to find the necessary technical, financial and policy measures to meet the large growth of energy demands in this region, while ensuring that the objectives of sustainable development can be met.

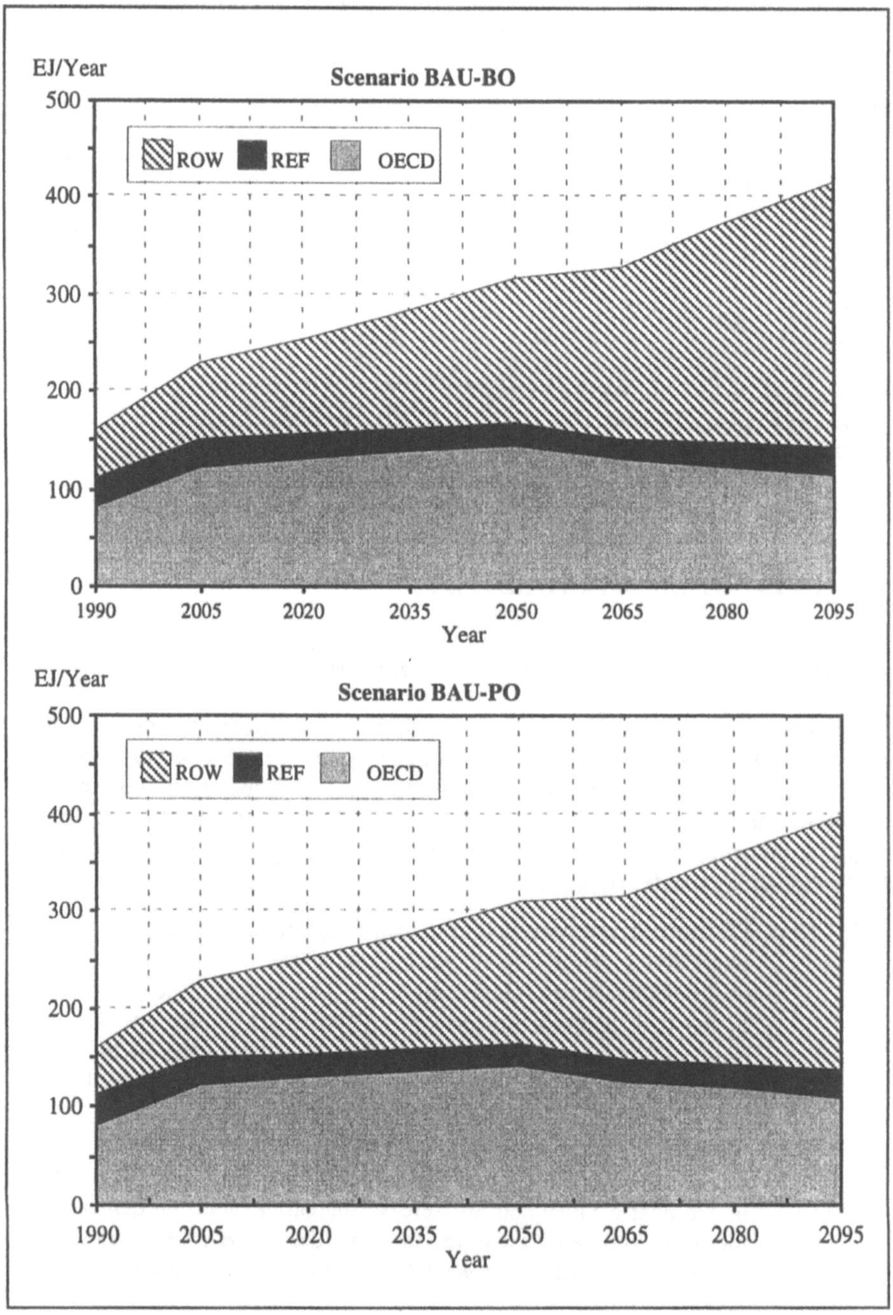

Figure 6a. Final energy demands by region – BAU scenarios

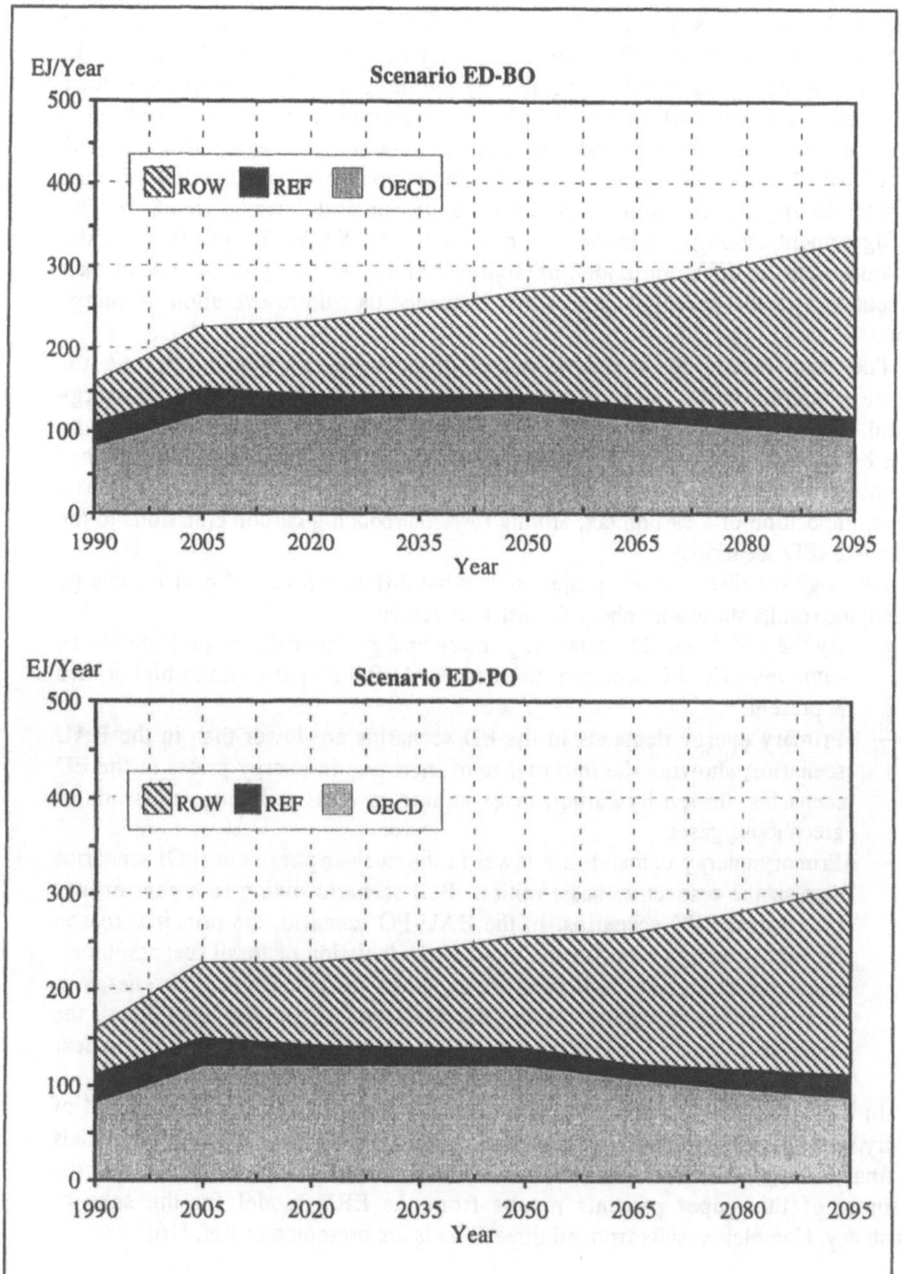

Figure 6b. Final energy demands by region – ED scenarios

The ERB results for FE demands, in each of the 13 regions treated by the model, were used as input to the GEM10R and LDNE21 models (with adaptation to the regional structure used in the respective model), and all three models produced estimates of primary energy (PE) demand. As may be seen in Figures 7a and 7b, the global primary energy demands calculated by the three modelling approaches for a given scenario differ by only some 20-25% from the lowest (LDNE21) to the highest (ERB) estimates. These PE demand estimates are in rather good agreement also with the IIASA/WEC Scenarios B and C, from which the economic growth (GDP) driving factors for energy demand were derived for the present study (see Section 3-4). This good agreement among the models, and with the IIASA/WEC results, provides confidence that the different modelling approaches are relatively consistent in their representation of the principal factors that influence the future evolution of energy demand.

The main reason for the variation in primary energy demand among the different scenarios is the price of energy, which has a feedback effect on energy demand (higher price leading to lower demand). Energy price increases may be driven by a number of factors, including: depletion of fossil fuel resources; forced phase-out of nuclear energy (PO scenarios) and replacement by higher cost energy sources; imposition of a carbon tax, aiming towards reducing carbon emissions to the atmosphere (ED scenarios).

Although the three models project somewhat different levels of primary energy demand, the results show a number of consistent trends:

- By the end of the 21st century, primary energy demands are projected to be some two (ED-PO scenario) to three (BAU-BO scenario) times higher than at present.
- Primary energy demands in the ED scenarios are lower than in the BAU scenarios, showing the importance of increases in energy prices in the ED scenarios, driven by carbon taxes aiming towards reducing emissions of greenhouse gases.
- Primary energy demands are lower in the nuclear phase out (PO) scenarios than in the respective basic option (BO) scenario owing to higher energy prices in the PO scenarios. In the BAU-PO scenario, the principal reason for higher energy prices is the more rapid depletion of fossil fuel resources, resulting in higher fossil fuel prices, when lower cost nuclear energy is eliminated as an option. In the ED-PO scenario, on the other hand, the increase in energy prices results mainly from the substitution of nuclear energy by higher cost renewable energy.

In light of the rather good agreement among the three models at the level of primary energy (PE), and taking into account that the GEM10R and LDNE21 models use final energy (FE) estimates from the ERB model as a starting point, the remainder of this paper presents results from the ERB model for the sake of consistency. Complete results from all three models are presented in Ref. [16].

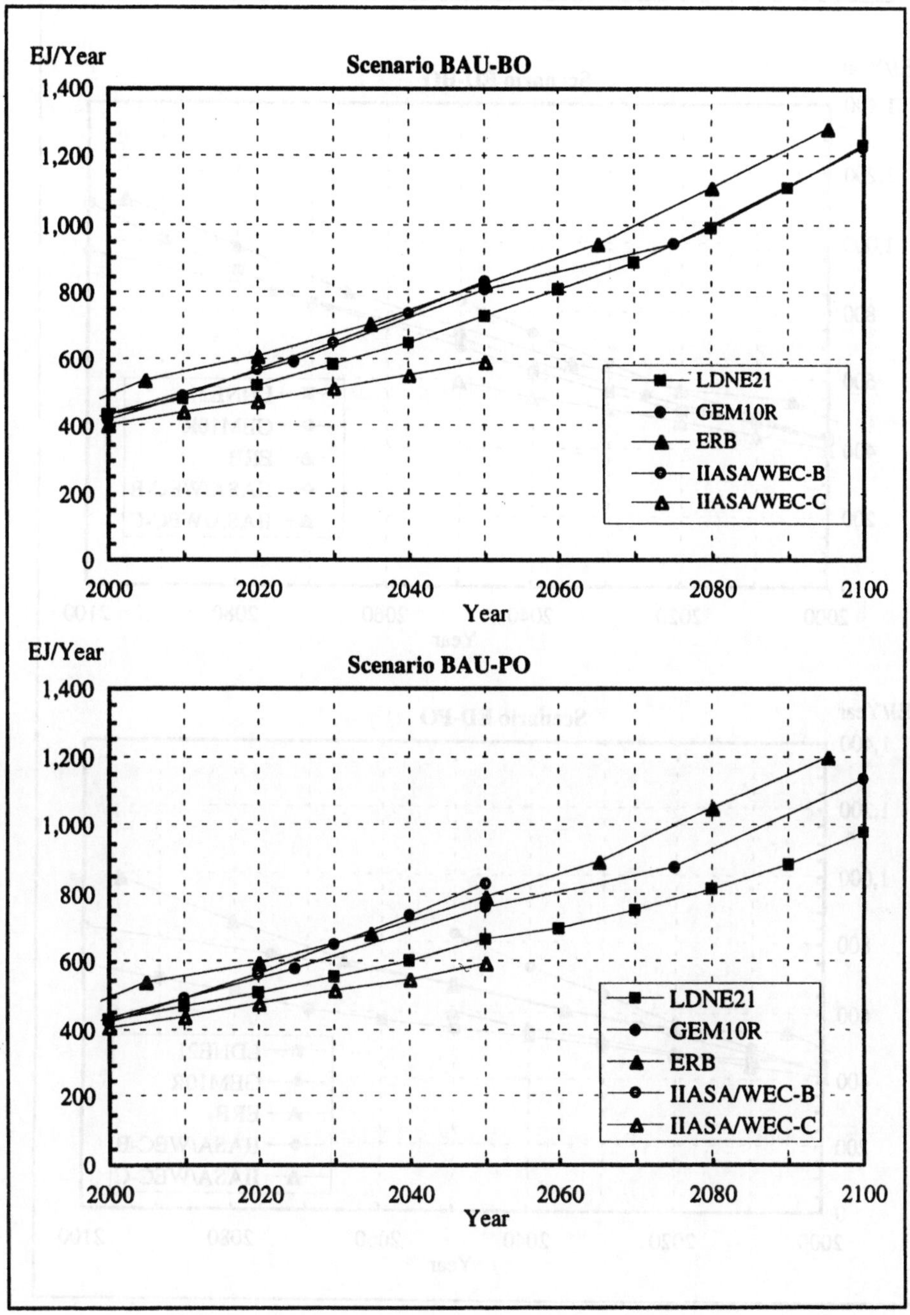

Figure 7a. Primary energy demand estimates – BAU scenarios

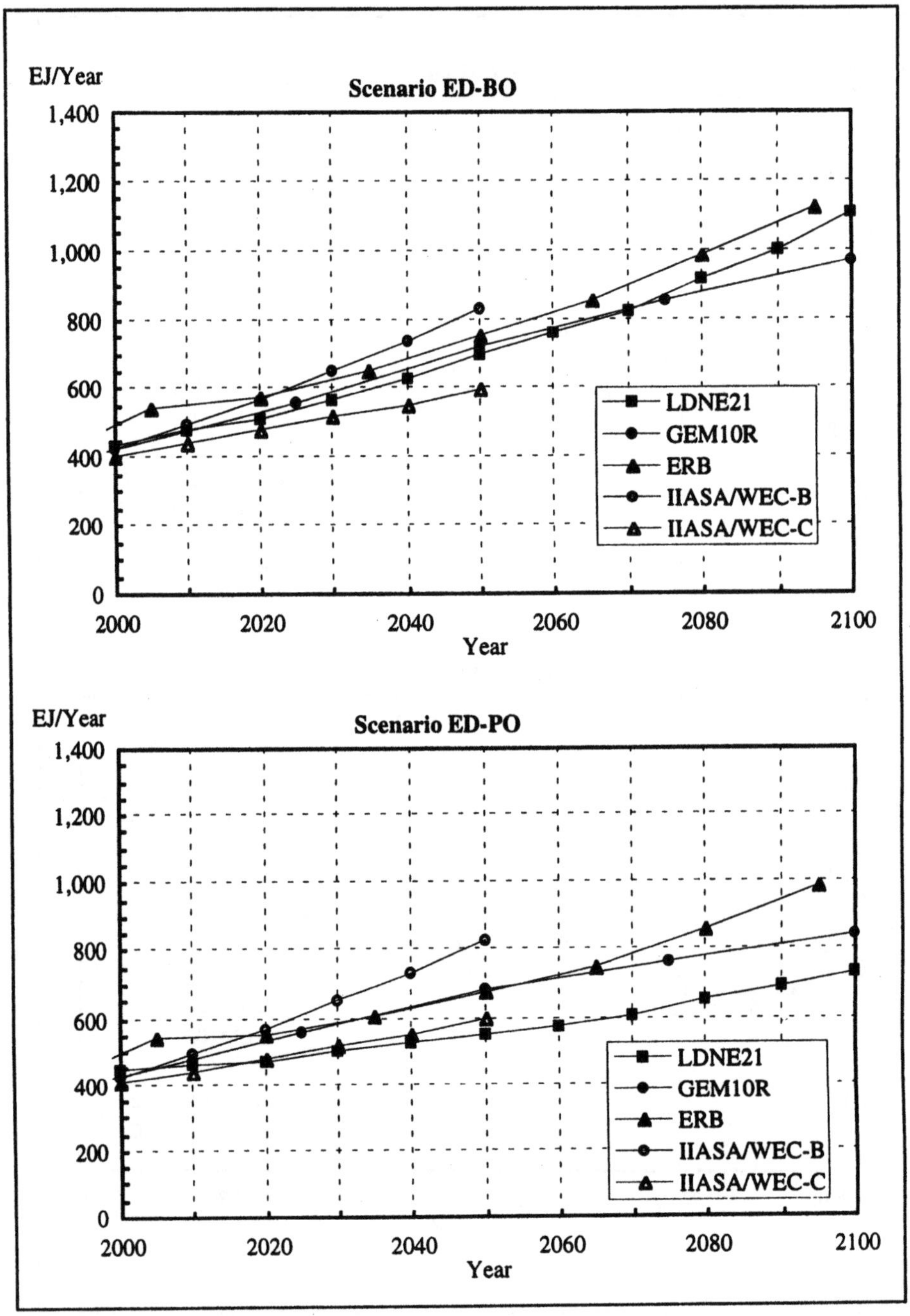

Figure 7b. Primary energy demand estimates – ED scenarios

The primary energy (PE) demands at the regional level (for the OECD, REF and ROW macro-regions used for reporting results in this study) are shown in Figures 8a and 8b. A number of general trends can be seen:

- *OECD region:* PE demands are estimated to increase from the present level of some 250 EJ, reaching a maximum of some 300-370 EJ[1] (depending on the scenario) by the middle of the next century, and decreasing to around 240-280 EJ by the year 2100.
- *REF region:* PE demands (about 80 EJ today) are projected to decrease up to the middle of the next century, reaching a minimum of some 60-70 EJ, and increasing thereafter to about 100-120 EJ by the year 2100.
- *ROW region:* in contrast to the other two regions, ROW shows a steady and large growth in PE demands throughout the next century, with PE demands increasing by a factor of 4 (scenario ED-PO) to 5 (scenario BAU-BO) by the end of the century.
- At present, the regional shares of global primary energy demand are about 53% in OECD, 15% in REF and 32% in ROW. By the year 2100, it is estimated that the shares will become around 25% in OECD, 9% in REF and 66% in ROW.
- Irrespective of uncertainties in the estimates, the general trend is clear - by around the middle of the next century, the primary energy consumption in the ROW region will be almost twice the combined consumption of the OECD and REF regions.

On a per capita basis, however, the regional distribution of primary energy demand shows a different picture (see Figures 9a and 9b). Although the total primary energy consumption in the ROW region at the end of the next century is estimated to be about 4 times higher than today's level, this energy will be needed by a population that will be around 1.8 times larger (see Figure 3) than at present. Therefore, the per capita energy consumption in the ROW region will only slightly more than double by the year 2100, and will remain far below that in either the OECD or REF region.

Figures 10a and 10b show the global levels of final energy (FE), secondary energy (SE) and primary energy (PE) demand for the four scenarios (ERB model results). At all three levels (FE, SE and PE) of the energy system, the two BAU scenarios show increasing energy demands. From the year 2000 to the year 2100, final energy demand about doubles; secondary energy demand grows by around 75%; and primary energy demand increases by a factor of about 2.5. In the two 'ecologically driven' (ED) scenarios, the increases in final and secondary energy demands are much less, with FE demand increasing by some 50%, SE demand increasing by only about 20-25%, and PE demand approximately doubling. As noted above, the lower energy demand in the ED scenarios, relative to the respective BAU scenario, is caused by higher energy price in the ED scenarios, which has a demand-reducing feedback effect through price-demand elasticity.

(1) EJ = Exa-Joules (10^{18} Joules)

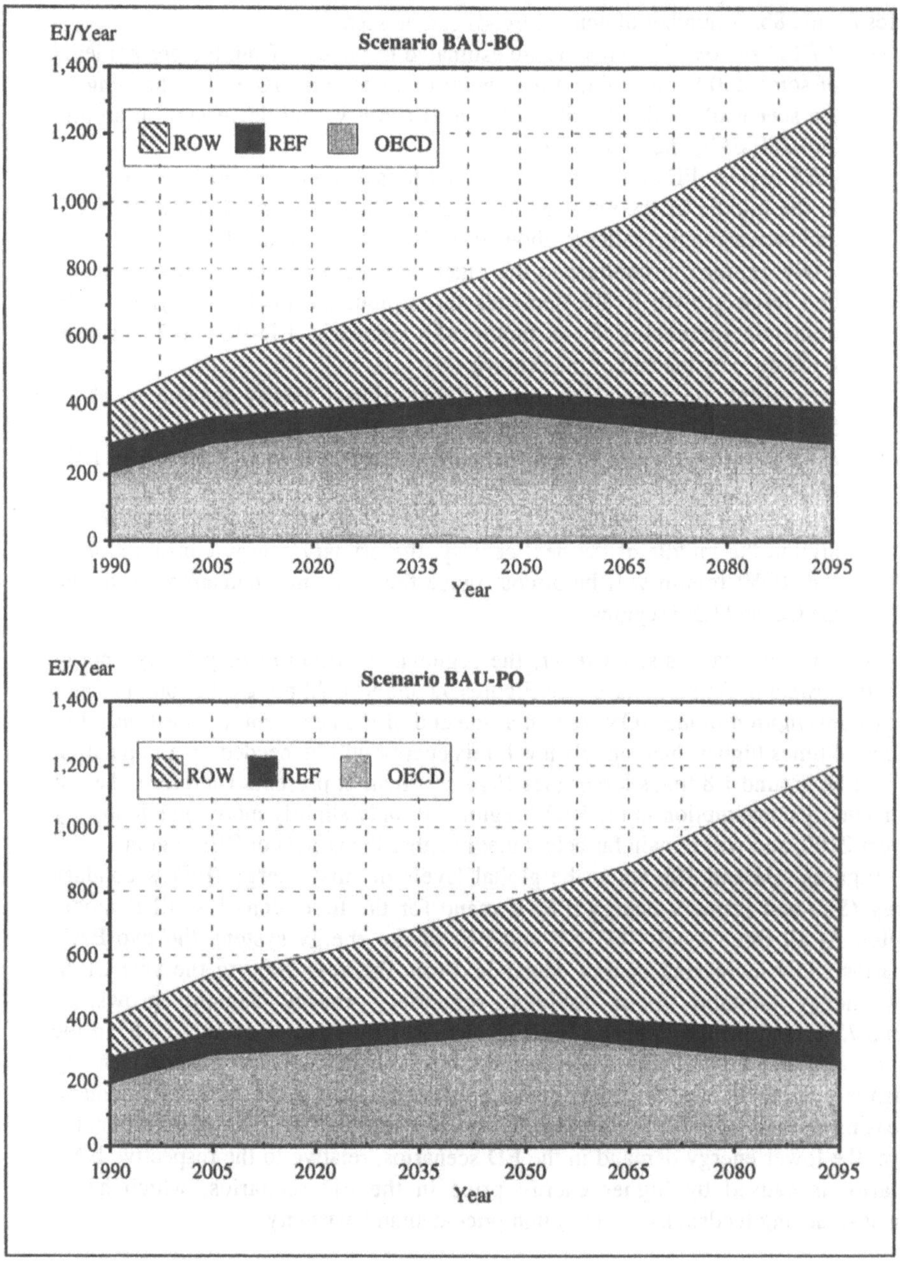

Figure 8a. Regional primary energy demands – BAU scenarios

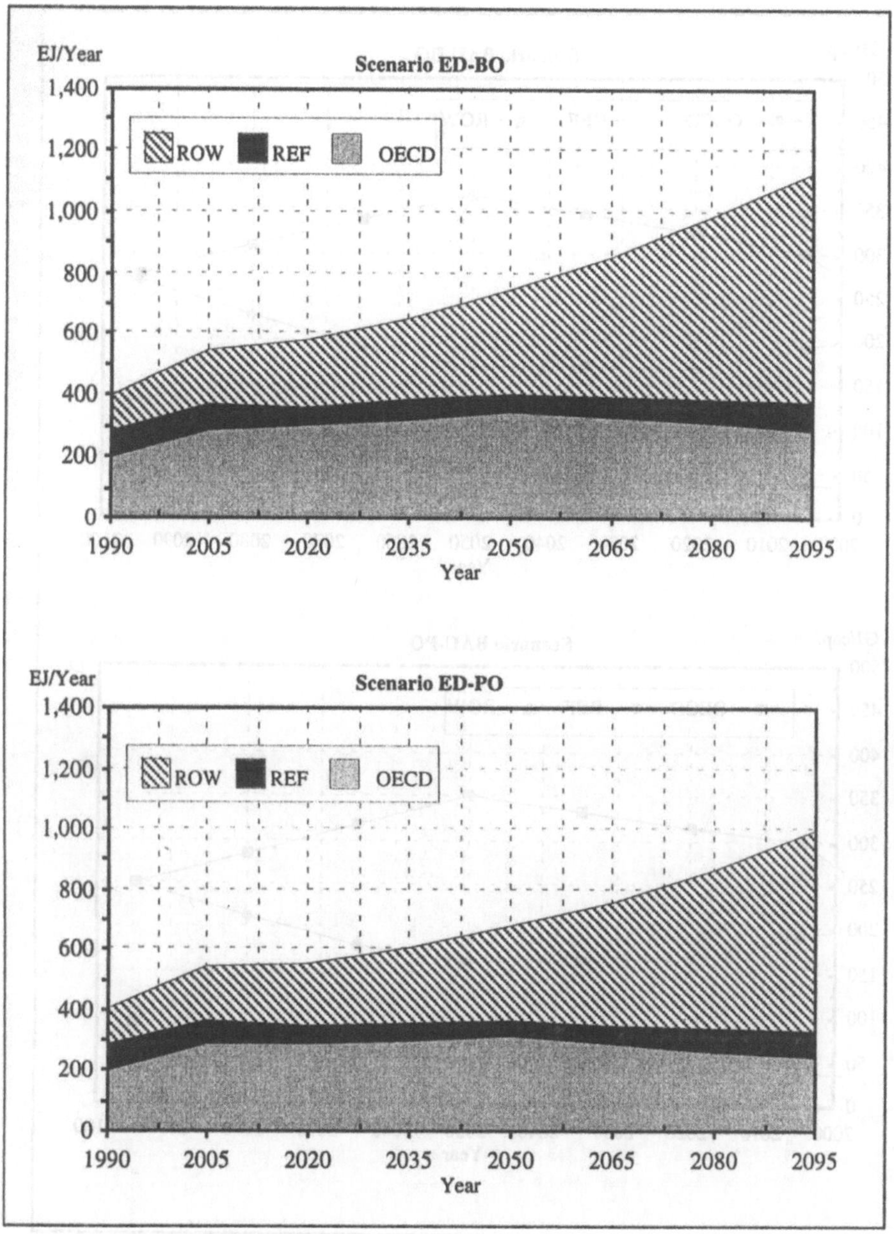

Figure 8b. Regional primary energy demands – ED scenarios

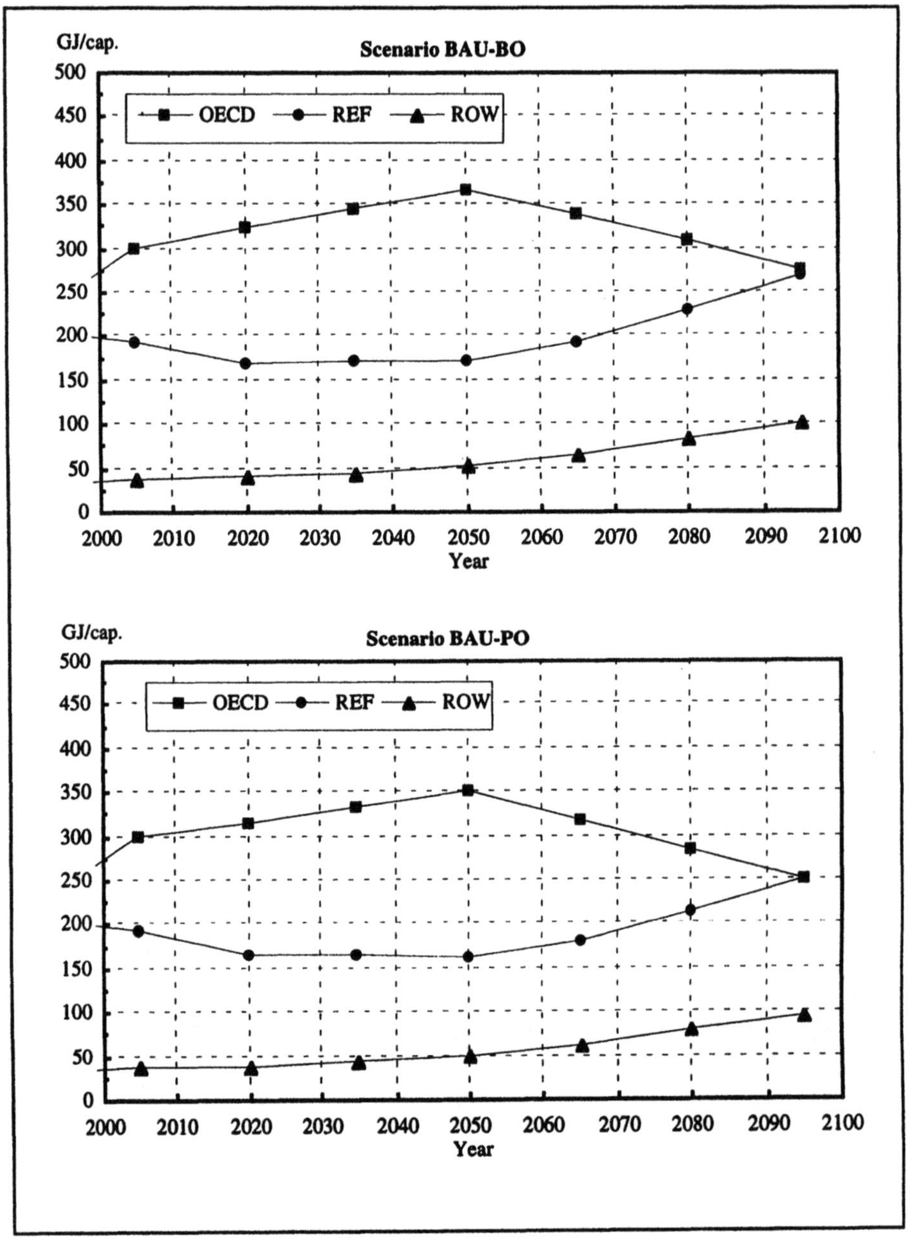

Figure 9a. Primary energy demand per capita – BAU scenarios

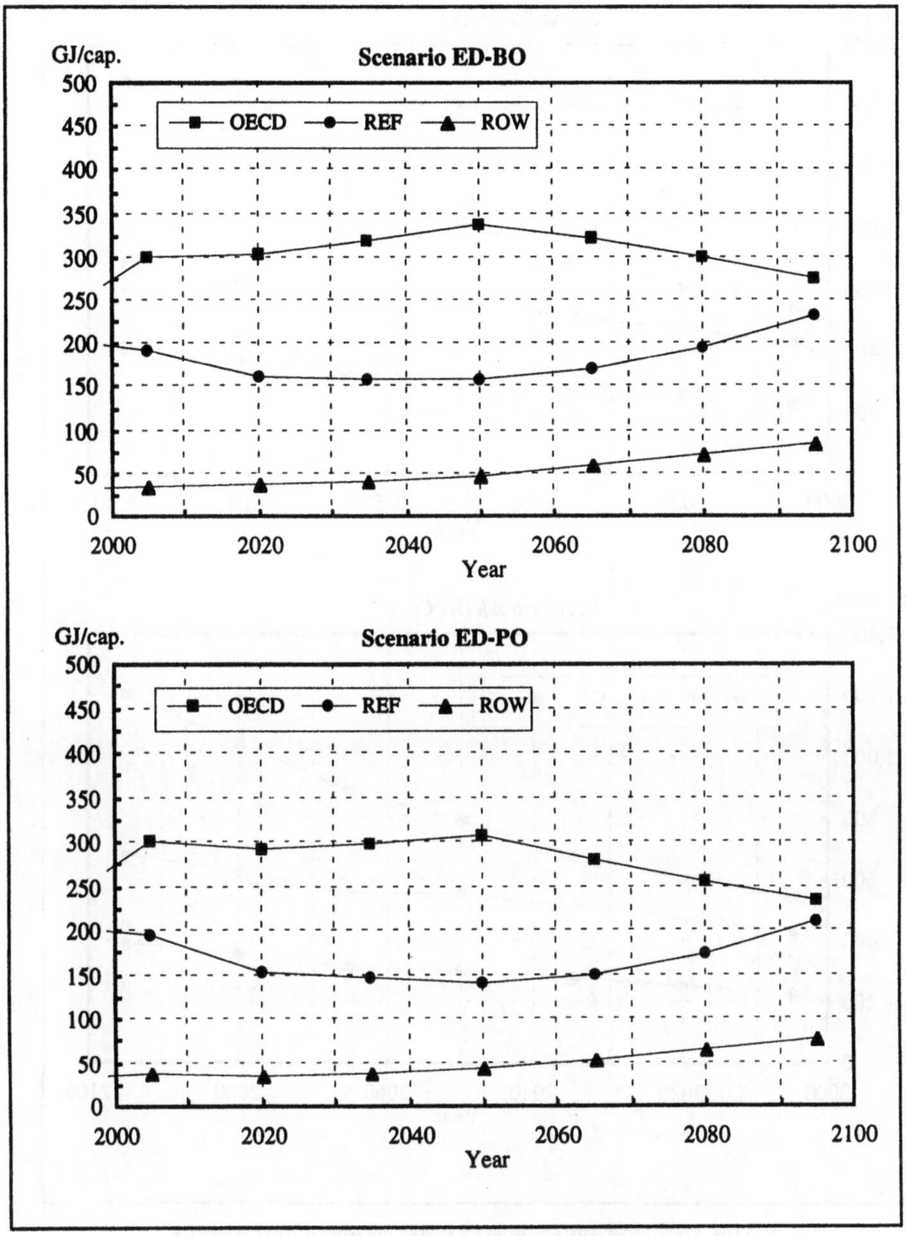

Figure 9b. Primary energy demand per capita – ED scenarios

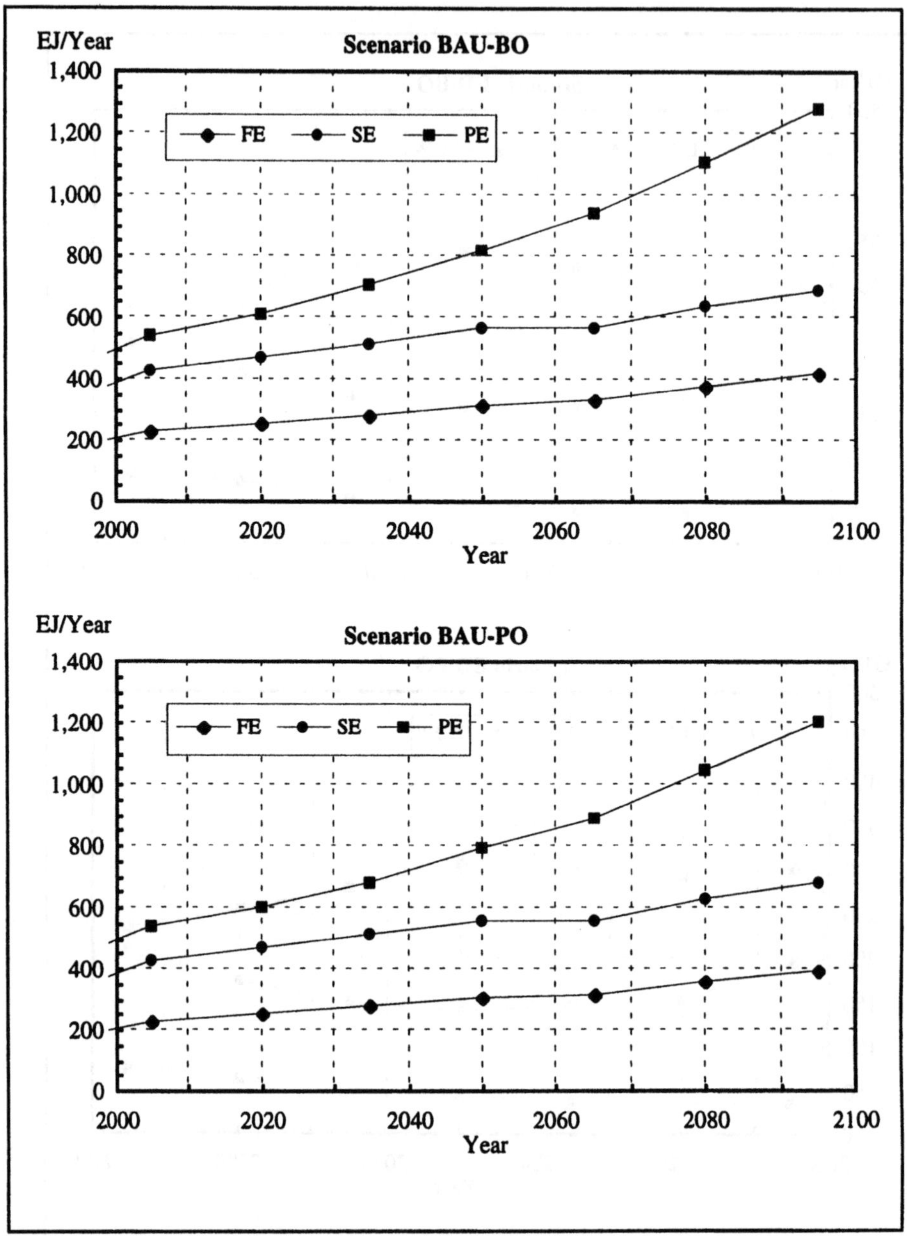

Figure 10a. Final, secondary and primary energy demands – BAU scenarios

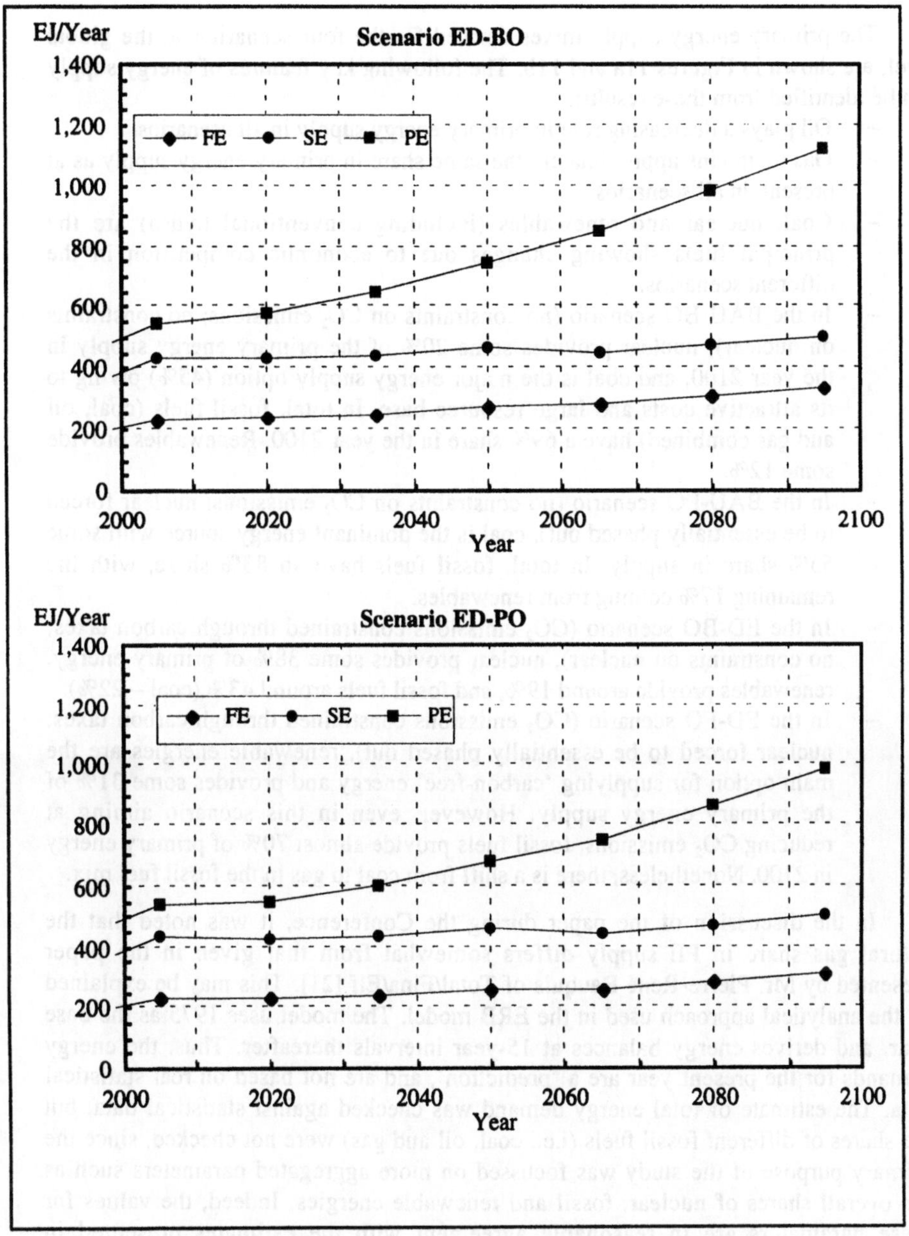

Figure 10b. Final, secondary and primary energy demands – ED scenarios

4.2 Primary Energy Supply Mix

The primary energy supply mixes derived for the four scenarios, at the global level, are shown in Figures 11a and 11b. The following key features of energy supply can be identified from these results:

- Oil plays a decreasing role in primary energy supply in all scenarios.
- Gas maintains approximately the same share in primary energy supply as at present, in all scenarios.
- Coal, nuclear and renewables (including conventional hydro) are the principal fuels showing changes due to economic competition in the different scenarios.
- In the BAU-BO scenario (no constraints on CO_2 emissions; no constraints on nuclear), nuclear provides some 20% of the primary energy supply in the year 2100, and coal is the major energy supply option (45%) owing to its attractive costs and large resource base. In total, fossil fuels (coal, oil and gas combined) have a 69% share in the year 2100. Renewables provide some 12%.
- In the BAU-PO scenario (no constraints on CO_2 emissions; nuclear forced to be essentially phased out), coal is the dominant energy source with some 55% share in supply. In total, fossil fuels have an 83% share, with the remaining 17% coming from renewables.
- In the ED-BO scenario (CO_2 emissions constrained through carbon taxes; no constraints on nuclear), nuclear provides some 38% of primary energy, renewables provide around 19%, and fossil fuels around 43% (coal – 22%).
- In the ED-PO scenario (CO_2 emissions constrained through carbon taxes; nuclear forced to be essentially phased out), renewable energies are the main option for supplying 'carbon-free' energy and provides some 31% of the primary energy supply. However, even in this scenario aiming at reducing CO_2 emissions, fossil fuels provide almost 70% of primary energy in 2100. Nonetheless, there is a shift from coal to gas in the fossil fuel mix.

In the discussion of the paper during the Conference, it was noted that the natural gas share in PE supply differs somewhat from that given in the paper presented by Mr. Pierre-René Bauquis of Total/Fina/Elf [21]. This may be explained by the analytical approach used in the ERB model. The model uses 1975 as the base year, and derives energy balances at 15-year intervals thereafter. Thus, the energy demands for the present year are a 'prediction', and are not based on real statistical data. The estimate of total energy demand was checked against statistical data, but the shares of different fossil fuels (i.e., coal, oil and gas) were not checked, since the primary purpose of the study was focussed on more aggregated parameters such as the overall shares of nuclear, fossil and renewable energies. Indeed, the values for these parameters are in reasonable agreement with the estimates presented in Reference 21.

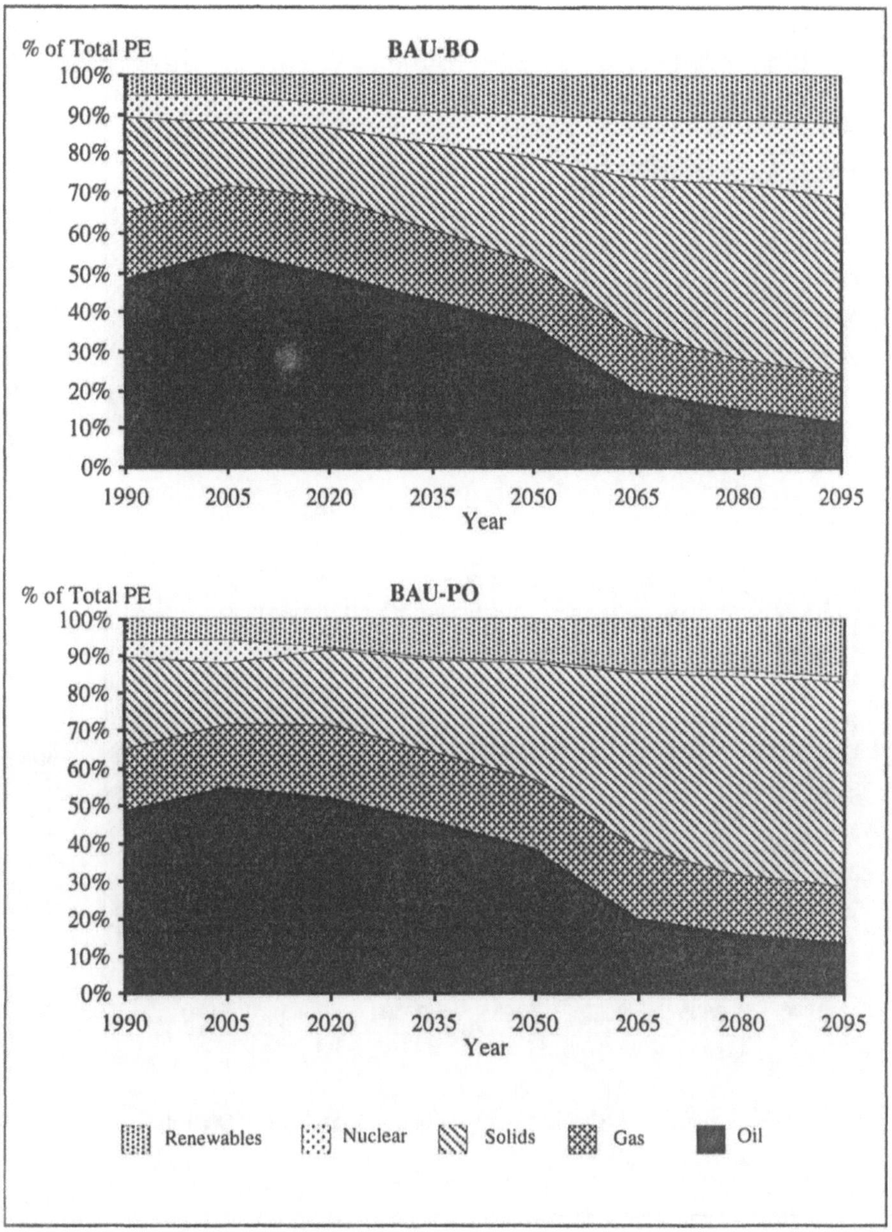

Figure 11a. Primary energy supply mixes – BAU scenarios

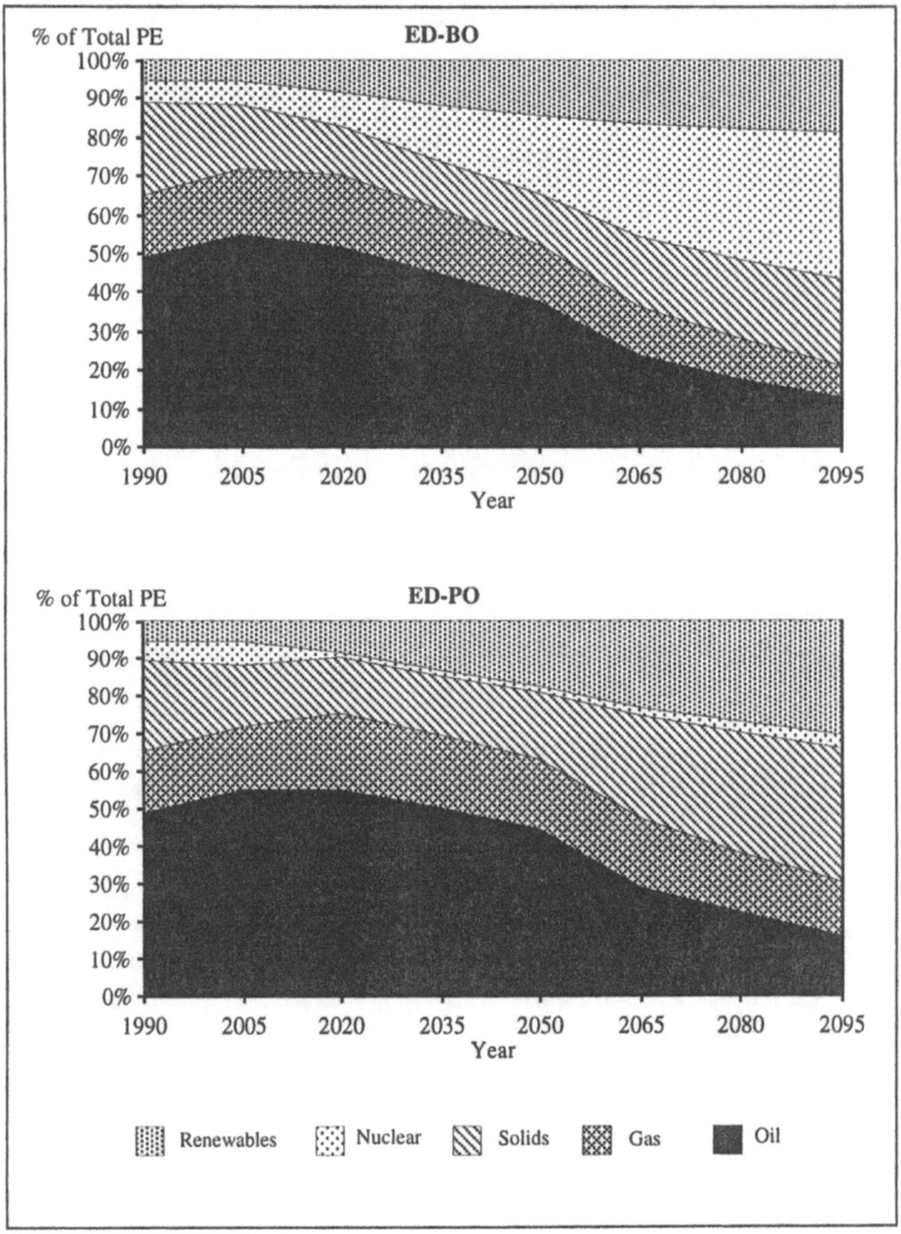

Figure 11b. Primary energy supply mixes – ED scenarios

4.3 Electricity Supply Mix

An important objective of the study was the analysis of the role of different energy sources (fossil, nuclear and renewable) in the long-term energy supply mix, under the assumptions taken in both the BAU scenarios (no constraints on CO_2 emissions) and the ED scenarios (aiming towards achieving stabilisation of atmospheric concentrations of CO_2). From the viewpoint of the IAEA and the NEA/OECD, there was particular interest in analysing the potential role of nuclear energy as a carbon-free electricity supply technology, and the degree to which nuclear energy could assist in reducing the costs of stabilising atmospheric concentrations of CO_2.

The global electricity generation mixes for each of the four scenarios are shown in Figures 12a and 12b. It can be noted that the small amounts of nuclear generation remain in the mix even in the 'nuclear phase-out' (PO) scenarios. In the ERB model, the capital cost of nuclear power plants was increased sharply in these scenarios, in an attempt to cause nuclear energy to disappear from the mix. However, the cost would have had to be increased even more dramatically in order to force nuclear completely from the mixture. In any case, the residual share of nuclear generation is essentially negligible.

A number of trends can be seen from the results:

- BAU-BO scenario: Nuclear energy captures an increasing share of electricity generation. In the year 2100, the mix is: nuclear, 35%: fossil fuels, 42% (coal, 23%); renewables (including conventional hydro, 23%.
- BAU-PO scenario: Nuclear essentially has been forced out of the mix, having only a residual share of 3% in the year 2100. Thus, the main supply options are fossil fuels (64%) and renewables (33%).
- ED-BO scenario: In this scenario, with constraints on CO_2 emissions and no constraints on nuclear expansion, nuclear penetrates more rapidly into electricity generation than in the BAU-BO scenario, reaching 54% in the year 2100. Fossil fuels are reduced to an 18% share and renewables contribute 27%. Both this scenario and the BAU-BO scenario (in both, nuclear is allowed to compete freely with other energy sources) show the favourable economics of nuclear electricity in comparison to electricity from 'new' renewable energy (i.e., renewable sources other than conventional hydro power).
- ED-PO scenario: With the forced phase-out of nuclear energy in this scenario, other than a remaining residual 6% share in the year 2100, renewable energies are the only 'carbon-free' energy source to be used for meeting the constraints on CO_2 emissions. Thus, renewable energies capture a large share of electricity generation, reaching around 55% by 2100. Nonetheless, fossil fuels still provide some 39% of electricity.

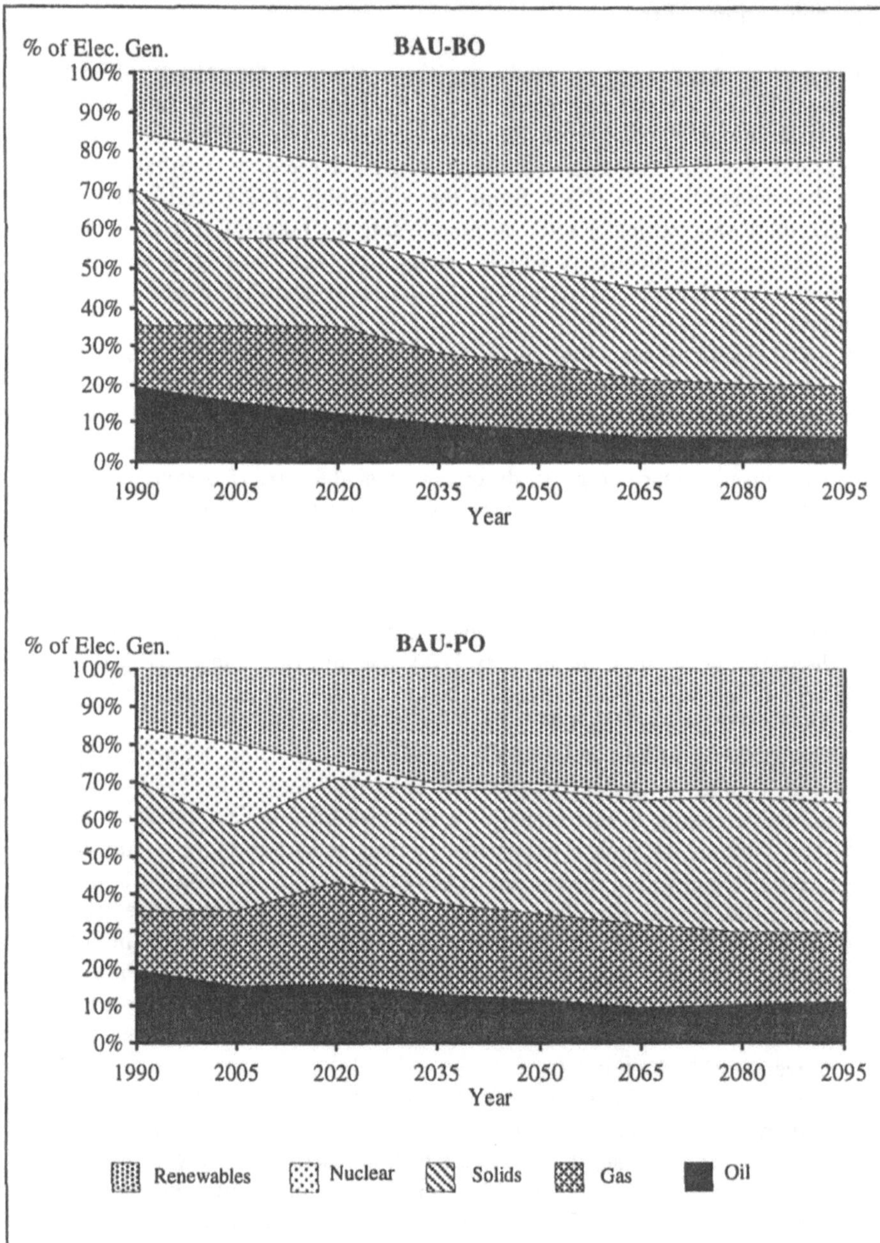

Figure 12a. Electricity generation mixes – BAU scenarios

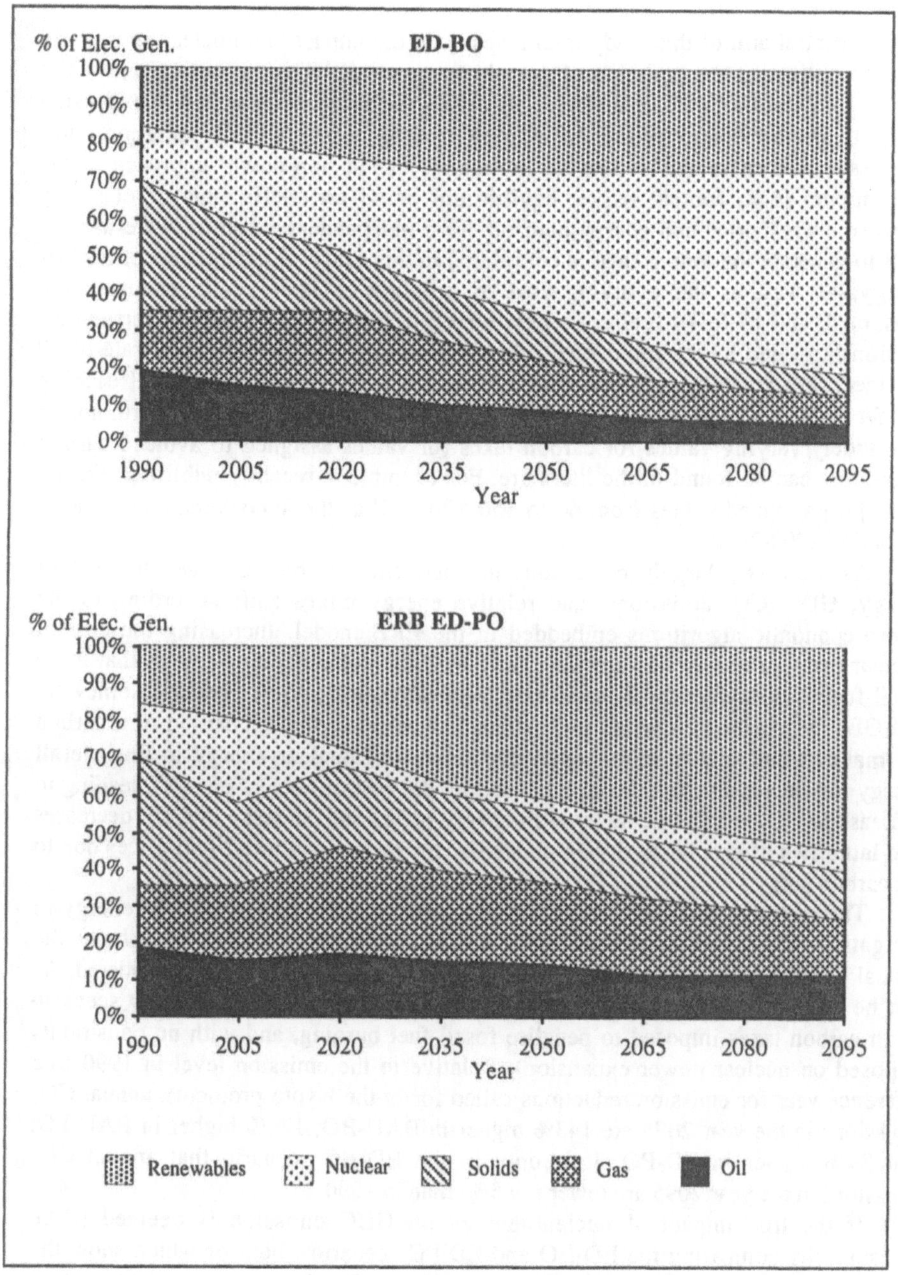

Figure 12b. Electricity generation mixes – ED scenarios

4.4 CO_2 Emissions

A central aim of this study is to investigate the impact that nuclear energy can have on mitigating the emission of greenhouse gases (GHGs), in particular of CO_2, and to compare the costs of different strategies (i.e., with nuclear power expansion and with nuclear energy phase-out) aiming towards mitigating CO_2 emission rates. This issue was examined within the limitations of the 'top-down' approach of the ERB model (e.g., forced market equilibrium based on coefficients of elasticity between energy price and energy demand. Two independent variables were used in ERB to generate the four scenarios considered in this study: a) capital cost of nuclear energy, as well as the evolving cost of uranium as lower cost resources are consumed, and b) fossil-fuel costs determined by fossil-fuel resources versus cost relationships, and by the imposition of carbon taxes at a step-wise linear rate of 30 $/tonneC (8.2 $/t CO_2) at intervals of 15 years, starting in the year 2005 [reaching 210 $/t C (57.3 $/t CO_2) in the year 2095. With regard to the latter, it is worth noting that widely varying values for carbon taxes (or values assigned to avoided carbon emissions) can be found in the literature. For example, a recently published French study [22] assigned values from 60 to 150 $/tonneC as the average carbon value in the period 2000-2050.

As these two largely exogenous independent variables change, the cost of energy, GDP, CO_2 emissions, and relative energy mixes shift according to the macro-economic algorithms embedded in the ERB model. Increasing the cost of nuclear energy decreases its market share, with the corresponding results that more fossil-fuel is consumed, CO_2 emissions increase, energy prices increase somewhat and GDP decreases by a small percentage. Likewise, as the imposition of a carbon tax makes fossil fuel more expensive, the share of nuclear energy in the overall energy mix increases (to an extent dictated by the cost of nuclear energy), leading to decrease in CO_2 emissions, increase in energy prices and decrease in GDP decreases (the latter two effects being driven mainly by the increase in fossil fuel prices due to the carbon tax).

The interplay between these variables, insofar as the role of nuclear energy in mitigating CO_2 emissions is concerned, is illustrated in Figure 13, which shows the annual global energy-related CO_2 emission rates for the four scenarios considered. As can be seen, CO_2 emissions increase in all scenarios except the ED-BO scenario (with carbon taxes imposed to penalise fossil fuel burning, and with no constraints imposed on nuclear power expansion). Relative to the emission level in 1990 (the reference year for emission reductions called for in the Kyoto protocol), annual CO_2 emissions in the year 2095 are 143% higher in BAU-BO; 174% higher in BAU-PO; and 34% higher in ED-PO. It is only in the ED-BO scenario that annual CO_2 emissions in the year 2095 are lower (by 5%) than in 1990.

If the true impact of nuclear energy on GHG emission is deemed to be measured by comparing the ED-BO and ED-PO scenarios, both of which show the effects of the carbon tax in reducing energy demand and in inducing shifts in fuel mixes (reducing fossil fuel shares), the results show that nuclear energy (in the ED-BO scenario) reduces CO_2 emissions by 29% in the year 2095 relative to the nuclear phase-out scenario (ED-BO). In short, nuclear energy can play a significant role in energy supply strategies aiming towards reducing energy-related CO_2 emissions.

It should be emphasised that the analysis presented in the above paragraphs is based on annual CO_2 emission rates, and not on atmospheric CO_2 accumulations and/or average global temperature rise. Even for the ED-BO scenario, atmospheric CO_2 concentrations and attendant temperature rises are still increasing in the year

2095, even though the CO_2 emission rate has been reduced below 1990 levels. Thus, if the aim of energy policies would be to stabilise atmospheric concentrations of CO_2, the emission rates would have to be reduced even further than in the ED-BO scenario.

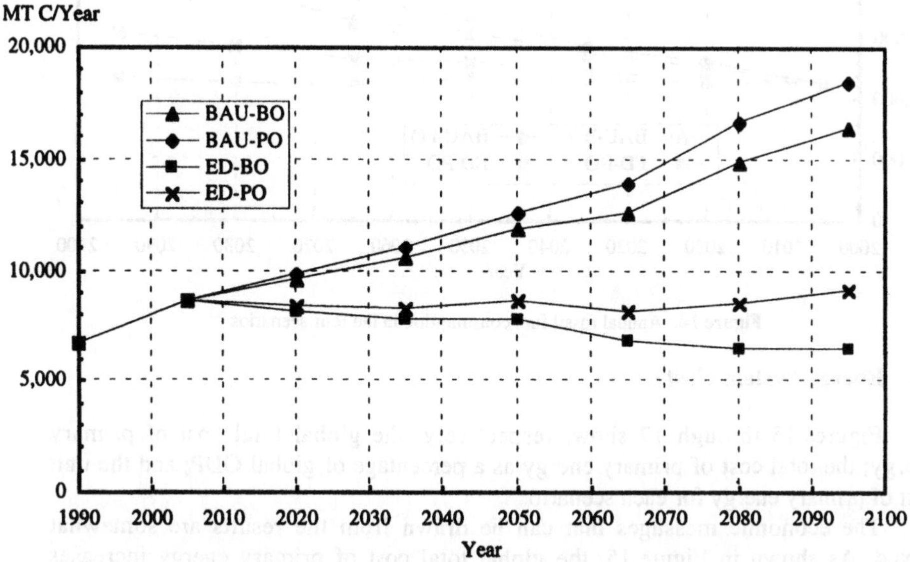

Figure 13. Carbon dioxide emissions (Megatonnes C/year

It should be emphasised that the analysis presented in the above paragraphs is based on annual CO_2 emission rates, and not on atmospheric CO_2 accumulations and/or average global temperature rise. Even for the ED-BO scenario, atmospheric CO_2 concentrations and attendant temperature rises are still increasing in the year 2095, even though the CO_2 emission rate has been reduced below 1990 levels. Thus, if the aim of energy policies would be to stabilise atmospheric concentrations of CO_2, the emission rates would have to be reduced even further than in the ED-BO scenario.

4.5 Fossil Fuel Savings

Figure 14 shows the total annual consumption of fossil fuels (coal, oil and gas), expressed in million tonnes of oil equivalent (MTOE). As may be seen, the scenarios that include nuclear (i.e., the BO scenarios) consume significantly less fossil fuels than do the nuclear phase-out (PO) scenarios. Also, it is apparent that the ED scenarios consume much less fossil fuels than do the BAU scenarios. This is not surprising, in light of the carbon tax that was imposed in the ED scenarios.

Comparing the two ED scenarios, the inclusion of nuclear power (ED-BO) leads to annual fossil fuel consumption being reduced by 25% in the year 2095, relative to the nuclear phase-out (PO) scenario. Thus, it is clear that nuclear power can contribute to significant savings in fossil fuel resources, while contributing also to reducing emissions of greenhouse gases.

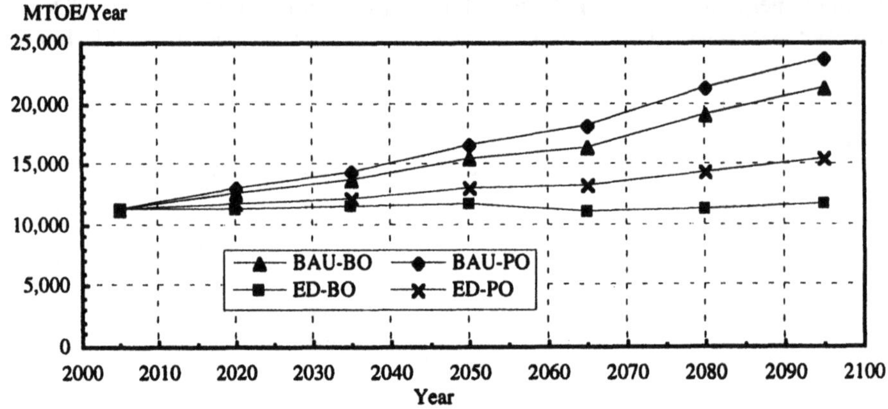

Figure 14. Annual fossil fuel consumption in the four scenarios

4.6 Energy System Cost

Figures 15 through 17 show, respectively: the global total cost of primary energy; the total cost of primary energy as a percentage of global GDP; and the unit cost of primary energy for each scenario

The economic messages that can be drawn from the results are somewhat mixed. As shown in Figure 15, the global total cost of primary energy increases continuously through the study period for all four scenarios, increasing by a factor of 3.8 (BAU-BO and ED-BO scenarios) to 5 (ED-PO scenario). As a percent of global GDP (see Figure 16), however, primary energy cost increases only by some 25% up to the middle of the next century and then decreases in the second half of the century. As shown in Figure 17, nuclear energy can have a major effect in reducing the unit costs of primary energy supply in the ED scenarios, that are aiming towards stabilising or reducing CO_2 emissions through the imposition of a carbon tax. As can be seen, the unit cost of primary energy in the ED-BO scenario is 40% below that in the ED-PO scenario in the year 2095.

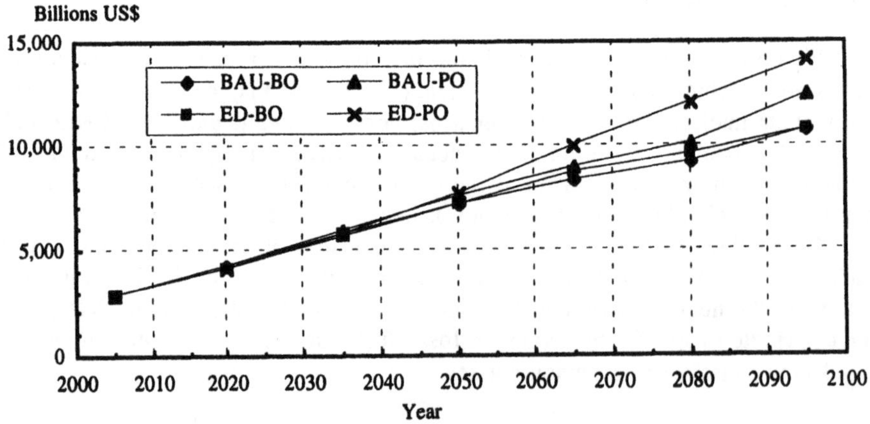

Figure 15. Global primary energy cost (billions US$)

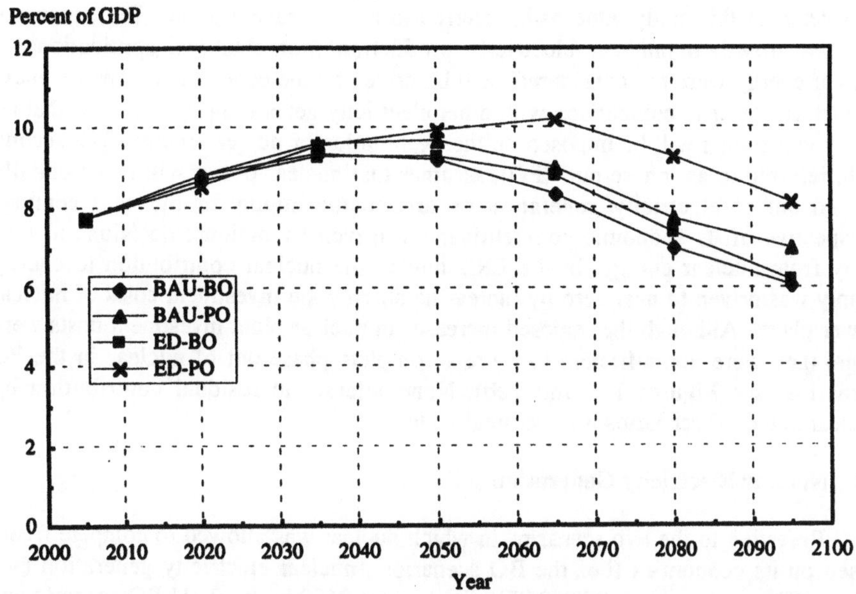

Figure 16. Primary energy cost as percent of global GDP

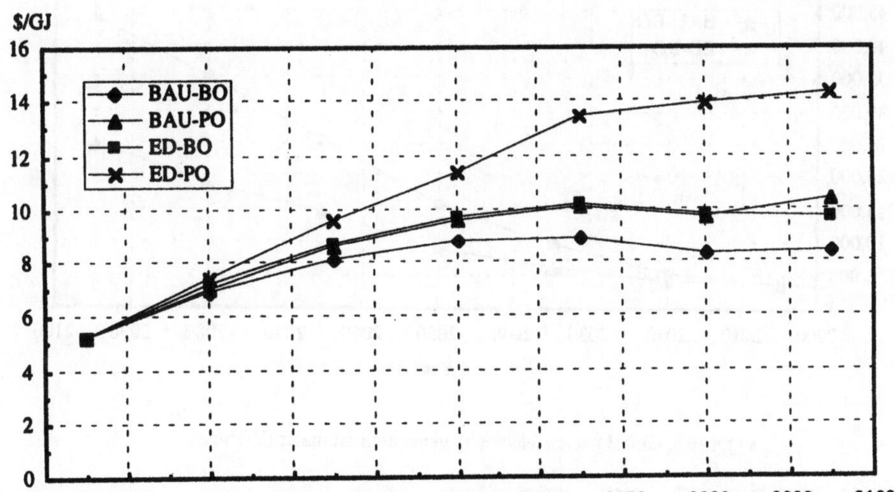

Figure 17. Unit cost of primary energy (US$/Giga-Joule)

5. ANALYSIS OF THE NUCLEAR DEVELOPMENT PATHS

As described in Section 3.3, two contrasting paths of nuclear development were considered in this study. One path, referred to as the 'basic option' (BO), assumes that the growth in nuclear electricity production (non-electrical applications of nuclear energy were not considered) will be driven by the economic competitiveness of nuclear power in comparison with other electricity generation options, and that no policy constraints will be imposed on the exploitation of nuclear energy. The second path, referred to as 'phase-out' (PO), assumes that nuclear power will be essentially phased out of electricity generation by around the middle of the next century, irrespective of its economic competitiveness, driven by national decisions to turn away from nuclear energy. In the ERB model, the nuclear contribution to energy supply was driven to near zero by increasing sharply the investment costs of nuclear power plants. Although the imposed increases in nuclear plant investment costs were large, they were not sufficient to force a complete phase-out of nuclear in the PO scenarios (see Figures 12a and 12b). Nonetheless, the residual contribution by nuclear in the PO scenarios is rather negligible.

5.1 Nuclear Electricity Generation

Referring to the two scenarios in which nuclear was allowed to compete freely based on its economics (i.e., the BO scenarios), nuclear electricity generation (see Figure 18) grows to some 25,000 TWh in the year 2100 in the BAU-BO scenario and to some 45,000 TWh in the ED-BO scenario. These levels of nuclear generation are some 9 to 15 times higher, respectively, than at present.

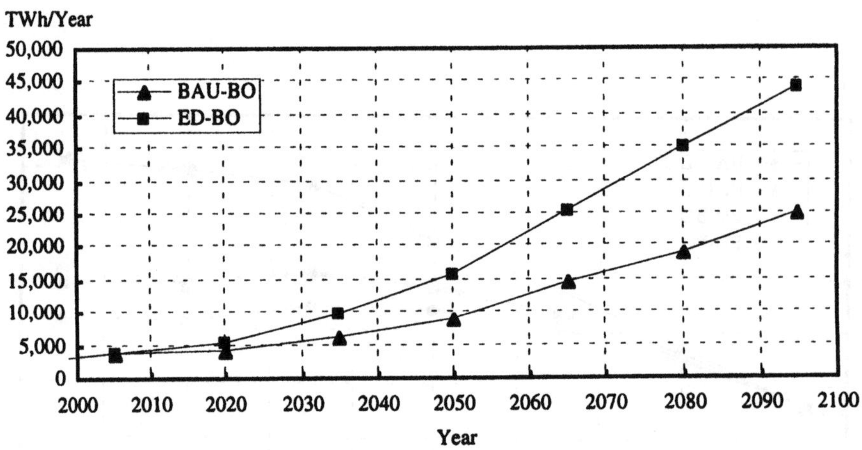

Figure 18. Global nuclear electricity generation estimates (TWh/year

5.2 Nuclear Power Capacity

The growth in nuclear power capacity was calculated from the nuclear electricity generation based on a plant load factor of 80%. This assumption on load factor is deemed to be conservatively low, and some nuclear plants operating today achieve higher load factors. As shown in Figure 19, the capacity of nuclear power plants in operation world-wide in the year 2100 reaches about 3,900 GWe in the BAU-BO scenario and about 6,700 GWe in the ED-BO scenario, or some 9 to 15 times, respectively, the capacity in operation as of January 2000.

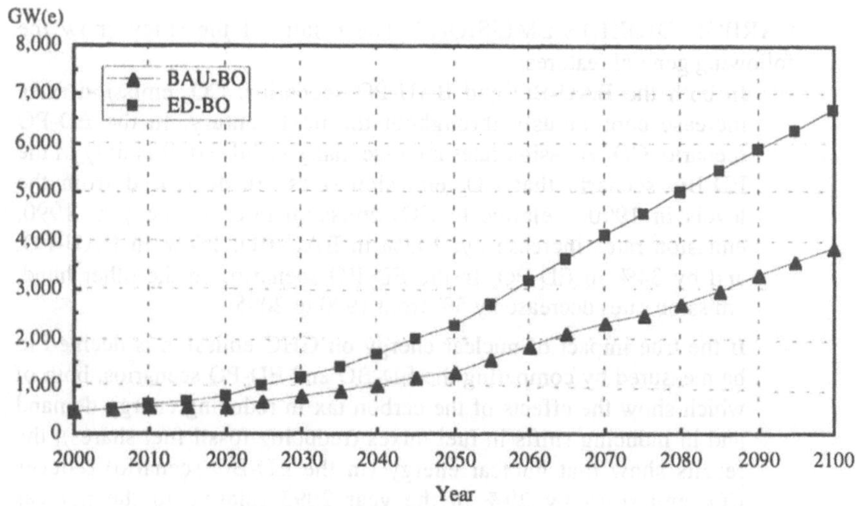

Figure 19. Global nuclear capacity in operation (GWe)

5.3 Annual Additions of Nuclear Power Capacity

Annual net additions of nuclear capacity were calculated from the nuclear capacity curves shown in Figure 19 above. As can be seen from Figure 20, the annual capacity additions reach a maximum of some 55 GWe in the BAU-BO scenario and around 95 GWe in the ED-BO scenario. Based on past experience, with some 40 GW of capacity having been added in some years, these levels of capacity addition are considered to be quite feasible for the nuclear industry, although requiring expansion beyond the nuclear plant manufacturing capability existing today.

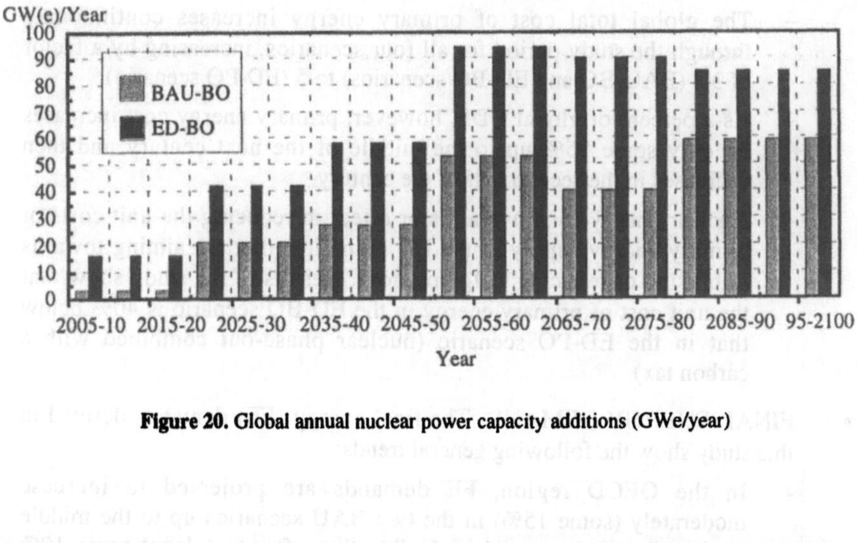

Figure 20. Global annual nuclear power capacity additions (GWe/year)

6. MAIN FINDINGS

- CARBON DIOXIDE EMISSIONS. The results of the study show the following general features:
  - In both the BAU-BO and BAU-PO scenarios, CO_2 emission rates increase continuously throughout the next century. In the ED-PO scenario, CO_2 emission rates are essentially stabilised. It is only in the ED-BO scenario that CO_2 emission rates are decreased from the levels in 1990. Relative to CO_2 emission rates in the year 1990, emission rates increase by: 143% in BAU-BO; 174% in BAU-PO; and by 34% in ED-PO. In the ED-BO scenario, on the other hand, emission rates decrease by 5% from 1990 to 2095.
  - If the true impact of nuclear energy on GHG emission is deemed to be measured by comparing the ED-BO and ED-PO scenarios, both of which show the effects of the carbon tax in reducing energy demand and in inducing shifts in fuel mixes (reducing fossil fuel shares), the results show that nuclear energy (in the ED-BO scenario) reduces CO_2 emissions by 29% in the year 2095 relative to the nuclear phase-out scenario (ED-PO). In short, nuclear energy can play a significant, although not dominant, role in CO_2 reduction strategies
  - It should be emphasised that the analysis presented in the above paragraphs is based on annual CO_2 emission rates, and not on atmospheric CO_2 accumulations and/or average global temperature rise. Even for the ED-BO scenario, atmospheric CO_2 concentrations and attendant temperature rises are still increasing in the year 2095, even though the CO_2 emission rate been reduced below 1990 levels. Thus, if the aim of energy policies would be to stabilise atmospheric concentrations of CO_2, the emission rates would have to be reduced below those shown for the ED-BO scenario.
- ENERGY SYSTEM COST. The economic messages that can be drawn from the results are somewhat mixed:
  - The global total cost of primary energy increases continuously through the study period for all four scenarios, increasing by a factor of 3.8 (BAU-BO and ED-BO scenarios) to 5 (ED-PO scenario).
  - As a percent of global GDP, however, primary energy cost increases by only some 25% up to the middle of the next century and then decreases in the second half of the century.
  - Nuclear energy can have a major effect in reducing the unit costs of primary energy supply in the ED scenarios, that are aiming towards stabilising or reducing CO_2 emissions. Results of the study show that the unit cost of primary energy in the ED-BO scenario is 40% below that in the ED-PO scenario (nuclear phase-out combined with a carbon tax).
- FINAL ENERGY DEMAND. The final energy (FE) demands derived in this study show the following general trends:
  - In the OECD region, FE demands are projected to increase moderately (some 15%) in the two BAU scenarios up to the middle of the next century, and to decline thereafter to a level some 10% below that in the year 2000. In the two 'ecologically driven' (ED)

scenarios, FE demands are projected to remain essentially stable up to the middle of the next century, and to decline thereafter to a level in the year 2100 that is 25%-30% below that in the year 2000.

- In the REF region, FE demands remain relatively stable up to the year 2100 in all scenarios, undergoing a small (some 10%) decrease up to the middle of the next century and an increase thereafter, with the demand in the year 2100 being about the same as (the two BAU scenarios), or slightly lower than (the two ED scenarios), the demand in the year 2000.
- In the ROW region, which encompasses the presently developing countries and which is the region of continuing population growth, FE demands are projected to increase by a factor of 3 to 4 by the year 2100. Therefore, the challenge in the next century will be to find the necessary technical, financial and policy measures to meet the large growth of energy demands in this region, while ensuring that the objectives of sustainable development can be met.

• PRIMARY ENERGY DEMAND. At the regional level (for the OECD, REF and ROW macro-regions used for reporting results in this study) a number of general trends can be seen:
  - *OECD region:* PE demands are estimated to increase from the present level of some 250 EJ, reaching a maximum of some 300-370 EJ(depending on the scenario) by the middle of the next century, and decreasing to around 240-280 EJ by the year 2100.
  - *REF region:* PE demands (about 80 EJ today) are projected to decrease up to the middle of the next century, reaching a minimum of some 60-70 EJ, and increasing thereafter to about 100-120 EJ by the year 2100.
  - *ROW region:* in contrast to the other two regions, ROW shows a steady and large growth in PE demands throughout the next century, with PE demands increasing by a factor of 4 (scenario ED-PO) to 5 (scenario BAU-BO) by the end of the century.
  - At present, the regional shares of global primary energy demand are about 53% in OECD, 15% in REF and 32% in ROW. By the year 2100, it is estimated that the shares will become around 25% in OECD, 9% in REF and 66% in ROW.
  - Irrespective of uncertainties in the estimates, the general trend is clear - by around the middle of the next century, the primary energy consumption in the ROW region will be almost twice the combined consumption of the OECD and REF regions.

• PRIMARY ENERGY SUPPLY MIX. The mix of fuels used to supply primary show the following general trends:
  - Oil plays a decreasing role in primary energy supply in all scenarios.
  - Gas maintains approximately the same share in primary energy supply as at present, in all scenarios.
  - Coal, nuclear and renewables (including conventional hydro) are the principal fuels showing changes due to economic competition in the different scenarios.

- In the BAU-BO scenario (no constraints on CO_2 emissions; no constraints on nuclear), nuclear provides some 20% of the primary energy supply in the year 2100, and coal is the major energy supply option (45%) owing to its attractive costs and large resource base. In total, fossil fuels (coal, oil and gas combined) have a 69% share in the year 2100. Renewables provide some 12%.
- In the BAU-PO scenario (no constraints on CO_2 emissions; nuclear forced to be essentially phased out), coal is the dominant energy source with some 55% share in supply. In total, fossil fuels have an 83% share, with the remaining 17% coming from renewables.
- In the ED-BO scenario (CO_2 emissions constrained through carbon taxes; no constraints on nuclear), nuclear provides some 38% of primary energy, renewables provide around 19%, and fossil fuels around 43% (coal – 22%).
- In the ED-PO scenario (CO_2 emissions constrained through carbon taxes; nuclear forced to be essentially phased out), renewable energies are the main option for supplying 'carbon-free' energy, providing some 31% of the primary energy supply. However, even in this scenario aiming at reducing CO_2 emissions, fossil fuels provide almost 70% of primary energy in 2100. Nonetheless, there is a shift from coal to gas in the fossil fuel mix.

7. CONCLUDING REMARKS

It is emphasised that the scenario results reported in this paper should not be viewed as predictive of how energy demand and supply will evolve in the long-term future. Rather, the study should be viewed as an exploration of the possible evolution of demand and supply paths, under the particular assumptions taken for the major driving factors (population growth, trends in GDP per capita, energy efficiency improvement, costs of energy conversion technologies, energy resource levels and price trends, etc.).

A particular feature of the study was the use of three different modelling approaches, including both 'top-down' econometric modelling (the ERB model) and 'bottom-up' technological modelling (the GEM10R and LDNE21 models). The study therefore was able to draw on the complementary features of the different modelling approaches, such as the feedback effects of energy price on energy demand that are modelled in the top-down approach and the more detailed representation of energy conversion technologies in the bottom-up approach.

As was noted in Section 4.1, the global primary energy demands calculated by the three modelling approaches were in rather good agreement for all four scenarios, differing by only some 20-25% for a given scenario. The PE demand estimates derived by this study are in rather good agreement also with the IIASA/WEC Scenarios B and C, from which the GDP per capita for the present study were taken. This good agreement among the models, and with the IIASA/WEC results, provides some degree of confidence that the different modelling approaches are relatively consistent in their representation of the principal factors that influence the future evolution of energy demand.

The energy conversion systems considered in the study are limited to those that are known today, and that already are in commercial use or are viewed as being on the threshold of significant market penetration. With this limitation, the study did not include technologies that are still at the stage of research into their technical feasibility (e.g., nuclear fusion, space based solar), although the experts participating in the study had, in some cases, performed other studies in which such systems were included. The rationale for limiting the present study to known technologies is that some major decisions on technical options for meeting environmental objectives will have to be taken in the next 10-20 years, and within that time frame the decisions cannot be based on technologies that are not rather well proven at present.

With regard to fossil fuels, the study assumed an 'orderly' increase in prices as resources are consumed, reflecting the progression from easily extractable resources to those that are more difficult and more costly to extract. However, as shown by the experience during the past year or so, oil and gas prices are rather volatile even under the present condition of adequate resource levels. Therefore, it might be expected that the price impacts of higher fossil fuel consumption will be felt well before approaching the limits of fossil fuel resources. Such effects were not examined in the study, however.

Finally, it has to be noted that the relative contribution of nuclear energy will be influenced rather strongly by the evolution of the costs of nuclear in comparison with other energy sources. In this regard, important variables are the capital cost of nuclear power plants, the capital costs of renewable energy systems, and the costs of fossil fuel resources. It would be useful to carry out further studies in order to investigate fully the sensitivity of the results to changes in these variables.

8. ANNEX

LIST OF EXPERTS PARTICIPATING IN THE STUDY

Name	Institute	Location
Belyaev, L. S.*	Energy Systems Institute	Russia
Bennett, L. L.	IAEA Consultant	France
Bertel, E.	NEA/OECD	France
Filippov, S. P.*	Energy Systems Institute	Russia
Fujino, J.*	University of Tokyo	Japan
Kagramanian, V	IAEA	Austria
Krakowski, R.*	Los Alamos National Laboratory	USA
Marchenko, O. V.*	Energy Systems Institute	Russia
Na, I.-G.	Korean Energy Economics Institute	Korea, Rep. of
Okabe, Y.	IAEA	Austria
Yamagi, K.*	University of Tokyo	Japan
Zaleski, C. P.	Université Paris Dauphine	France

* Energy scenario modelling expert

9. REFERENCES

[1] UNITED NATIONS STATISTICAL DIVISION, Energy Statistics Data Base (1993), UN, New York (1995).

[2] WORLD ENERGY COUNCIL AND INTERNATIONAL INSTITUTE FOR APPLIED SYSTEMS ANALYSIS, Global Energy Perspectives to 2050 and Beyond, World Energy Council, London (1995).

[3] WORLD COMMISSION ON ENVIRONMENT AND DEVELOPMENT (Brundtland, G.H., Ch.), Our Common Future, Oxford University Press, Oxford, England (1987).

[4] UNITED NATIONS CONFERENCE ON ENVIRONMENT AND DEVELOPMENT, Agenda 21 - Action Plan for the Next Century, United Nations, New York (1992).

[5] INTERGOVERNMENTAL PANEL ON CLIMATE CHANGE (IPCC), Climate Change 1995 – Impacts, Adaptations and Mitigation of Climate Change: Scientific and Technical Analyses, Cambridge University Press, New York (1996).

[6] UNITED NATIONS, Kyoto Protocol to the United Nations Framework Convention on Climate Change, FCCC/CP/1997/L.7/Add.1 (10 December 1997), Kyoto (1997).

[7] WORLD BANK, Environmental Assessment Source Book, Technical Paper No. 140, Washington, D.C. (1991).

[8] WORLD BANK, GTZ, ÖKO-INSTITUTE, The Environmental Manual for Power Development, An Introduction to the EM Version 1.0, Darmstatt (1995).

[9] EUROPEAN COMMISSION, ExternE: Externalities of Energy, vol. 1, Summary, Report No. EUR 16520 EN, EC/DG-XII, Luxembourg (1995).

[10] INTERNATIONAL ATOMIC ENERGY AGENCY, Format and Structure of a Database on Health and Environmental Impacts of Different Energy Systems for Electricity Generation, IAEA-TECDOC-645, Vienna (April 1992).

[11] INTERNATIONAL ATOMIC ENERGY AGENCY, Methods for Comparative Risk Assessment of Different Energy Sources, IAEA-TECDOC-671, Vienna (October 1992).

[12] INTERNATIONAL ATOMIC ENERGY AGENCY, Health and Environmental Impacts of Electricity Generating Systems: Procedures for Comparative Assessment, Technical Reports Series, STI/DOC/010/394,Vienna (1999).

[13] INTERNATIONAL ATOMIC ENERGY AGENCY AND NUCLEAR ENERGY AGENCY (OECD), Nuclear Power: An Overview in the Context of Alleviating Greenhouse Gas Emissions, IAEA-TECDOC-793, Vienna (1995).

[14] VLADU, I.F., "Energy Chain Analysis for Comparative Assessment in the Power Sector", in Electricity, Health and the Environment: Comparative Assessment in Support of Decision Making (Proc. Int. Symp., Vienna, 16-19 October 1995), IAEA Proceedings Series, STI/PUB/975, IAEA, Vienna (1996).

[15] INTERNATIONAL ATOMIC ENERGY AGENCY, Enhanced Electricity System Analysis for Decision Making – A Reference Book, DECADES Project Document No. 4, Vienna (June 2000).

[16] INTERNATIONAL ATOMIC ENERGY AGENCY AND NUCLEAR ENERGY AGENCY OF OECD, Scenarios of Nuclear Power Growth in the 21st Century, Report of an Expert Group Study, unpublished report, IAEA, Vienna (23 July 1999).

[17] ROGNER, H.-H., "An Assessment of World Hydrocarbon Resources", Annual Review of Energy and the Environment, 22:217-262, Annual Reviews Inc. (1997).

[18] NUCLEAR ENERGY AGENCY OF OECD AND INTERNATIONAL ATOMIC ENERGY AGENCY, Uranium 1997: Resources, Production and Demand, OECD, Paris (1998).

[19] GRINBLAT, J.-A., Chief, Population Estimates and Projections Section, Population Division, Department for Economic and Social Information and Policy Analysis, United Nations Secretariat, New York (personal communication, 23 April 1998).

[20] NUCLEAR ENERGY AGENCY OF OECD AND INTERNATIONAL ENERGY AGENCY, Projected Costs of Generating Electricity (Update 1998), OECD, Paris (1998).

[21] BAUQUIS, P.-R., Constraints on fossil fuels supplies for the next half century, Total/Fina/Elf, paper presented in the present conference.

[22] CHARPIN, J.-M., DESSUS, B. AND PELLAT, R., Étude économique prospective de la filière électrique nucléaire (An economic assessment of the nuclear power industry), Paris, July 2000.

NUCLEAR PLANT FINANCIAL PERFORMANCE IN A RESTRUCTURED UTILITY SYSTEM

Shelby T. Brewer[1]

1. INTRODUCTION

In a paper[2] two years ago to this forum (in Paris), I outlined the U.S. transition to an economically deregulated power generation economy, and what this could mean for nuclear power. My conclusion was then and it continues to be that this (market-driven economic competition) does not mean the end of nuclear power. For new nuclear plant construction starts, in, say, the next five years, the transition does not auger well, because it will be problematic for new nuclear plants to compete with natural gas-fired generation. However, the acquisition transactions in the population of existing, operating nuclear plants, offers a huge capital formation opportunity to rekindle nuclear power, this time as a market-driven technology, sans the nanny-coddling aspects of the first nuclear era. That is because nuclear plants, which are economic now, will be even more so in a market-driven economy. Earnings will not be constrained artificially as in the present regulated monopoly structure. Very large returns to equity can be realized, accrued and applied (later) to construction of new standardized, pre-licensed plants, and to other nuclear programs, such as the breeder.

My purpose today is to concentrate on the practical financial nuts-and-bolts, from the perspective of a party acquiring US nuclear power plants (NPPs) during this transition. More specifically, I will present a financial model (NPPACQ) for evaluating acquisition candidates, and give some examples, and some parametrics.

---

[1] President and Chief Executive Officer, Commodore Nuclear, Alexandria, VA

[2] *Nuclear Power and the U.S. Transition to a Restructured, Competitive Power Generation Sector,* S.T. Brewer, October 1998, Paris, France.

Global Warming and Energy Policy, Edited by Kursunoglu *et al.*
Kluwer Academic/Plenum Publishers, New York 2001

2. BASIC PREMISES - INVESTMENT RETURN EVALUATION

The financial model, NPPACQ©, was developed from the perspective of a potential purchaser of an operating nuclear plant, a purchaser who obviously needs to (a) compete with other power generators, while (b) maximizing his return on *equity* investment. These objectives are in stark contrast with the old regulated monopoly paradigm, which the new deregulation structure will replace over time. In the regulated monopoly format, a principal driver of utilities is to gain state Public Utilities Commissions (PUCs) approval for loading costs into rates (c/kwh) charged to electricity consumers. Return on capital (particularly equity capital), constrained to certain limits in the regulated regime, is lumped in the utility mind-set with debt capital, and treated like another *cost* element, like fuel, operating and maintenance, and so on. In the "old regime", there is little or no incentive to enhance profit (equity return), as it is regulated, and like all other costs are passed on to the consumer, in this fatal political embrace between utility bureaucracies and the state PUC bureaucracies.

Thus, the model presented here reflects two fundamental differences with the current vertically integrated regulated monopoly structure: (a) plants must compete in an unregulated power generation market; and (b) equity return is an objective financial objective.

Therefore, NPPAQC is a ***cash flow*** model, with cash flows from and to equity investors being the central focus. In the NPPACQ model depreciation (an accrual accounting abstraction) enters the computation only as a tax matter: depreciation of total capital is income tax-deductible. Otherwise, *depreciation* has no real and relevant meaning, except for the phony reporting each publicly traded company makes each year. *Income tax is treated as a cash flow, and depreciation is relevant only to income tax computations.*

Similarly, *debt service* is treated as a cash flow, another distinction between the current and future regimes. NPPACQ uses the uniform payment formula in MS-Excel. *The interest component of this cash flow is tax deductible.* Since interest on debt is a real phenomena both as a cash out flow and a tax phenomena, that enters the model.

Contributions to a decommissioning and decontamination fund (D&D fund) are treated as cash flows. The current balance is deducted from the estimated cost at end of life and an annual contribution is computed for the new owner. It is held in escrow.

3. OVERALL MODEL FORMULATION: ANNUAL EQUITY RETURN

Net after-tax cash flow to equity is computed for each year of remaining plant service:

ENCF = R – C – CapEx – DS – IT
where

ENFC = equity net cash flow for a given year
R= revenue for a given year
C= F .... Fuel cost
+OM ....operating and maintenance cost
+ FWF .... federal waste fee
+ DD .... payments to decommissioning/decontamination sinking fund
+ PT .... property tax
+ GA .... general and administrative costs

CapEx = annual capital expense outlay (equity component)
DS = total debt service payment, including principal and interest, on both original acquisition outlay and sum of debt component of CapEx's of preceding years.
IT = income tax

Income tax is given by

$IT = \tau[R\text{-}C\text{-}D\text{-}I]$

where

τ = composite federal and local income tax rate
D = total depreciation (debt & equity) in a given year
= depreciation on initial acquisition cost
$+ \sum[\textit{depreciation}, \textit{CapEx's}]$

I = interest in the given year on initial acquisition debt component
$+ \sum[\text{int}\,\textit{erest}, \textit{debtcomponents}, \textit{CapEx's}]$

The effective return on equity capital (IRR in the Excel software) is the solution for r in the equation

$$\sum_j [\textit{AnnualCashFlow}]j/[1+r]^j = 0.$$

3.1 Acquisition Price and Book Value

In the population of US operating NPPs, many are competitive with alternative power generation, or "market" power prices. NPPs which cannot compete economically with market, we consider "stranded" assets. Natural gas generating capacity is assumed to be the benchmark for competitive generation. It is assumed that a perspective buyer would not purchase a plant for a price above its market value.

In some states, retail deregulation legislation provides for the recover of stranded assets through a mechanism called "securitization". This means that the incumbent utility (seller) is made whole for the difference between the plant's book value and its market value, and thus the plant is saleable.

The NPPACQ model allows the purchaser to buy the plant at any price (above, at, or below market capital cost). Price of electricity (c/kwh) charged by the plant is another input variable, and a rationale assumption is to place at or below market.

Book value is the original capital cost as depreciated, plus any capital improvements (CAPEX) over the operating years, as depreciated.

Fuel inventory is another element of acquisition price. It is assumed to be two annual reloads. In some cases fuel inventory cost is greater than the plant acquisition cost, which seems counterintuitive, since nuclear capital costs [for a new plant coming on line] are typically higher than its alternatives, while its fuel costs are lower. The transaction involving TMI-1 is an example. It was sold for approximately $100 million--$25 million for the plant and $75 million for the fuel inventory. Incidentally, the TMI-1 book value at the time was several-fold above this number, suggesting that either there was a large dose of securitization or a large itch by the owner to divest.

Fuel inventory acquisition cost is added to plant acquisition price to determine capital required to purchase the plant. The model also provides for inclusion of other transaction costs, such as legal fees, and the like.

3.2 Debt and Debt Service

NPPACQ allows the user to input the debt-equity split for both a) the original acquisition cost and b) annual CAPEX outlays. The two leverage factors could be different. Debt service as treated as an annual cash out figure.

The Excel PMT algorithm is used for debt service, which is a uniform payment formula. For a) the original acquisition cost, the debt is amortized over the remaining life of the plant. For b) the debt service associated with annual CAPEX, the debt is amortized over the period beginning when the CAPEX occurs and the end of life of the plant.

3.3 Income Taxes

Income taxes are computed for each year of remaining operating life. For a given year, interest on debt and total depreciation (debt and equity) are tax deductible, along with fuel, operating and maintenance costs, G&A expenses, federal waste fund payments, payments to the decommissioning and decontamination (D&D) fund, etc. For a given year, income tax is the last computation made, to deduct from revenues and other costs, to yield return to equity.

3.4 Revenue

Revenue by year is computed simply by the product of price (c/kwh), capacity factor (%), and plant capacity (Mwe), and the appropriate conversion factor. The user can input each price and capacity factor, for each year. These input parameters can be tailored to power purchase agreements sought or obtained in the Acquisition Agreement, or by ceilings in state or regional law.

3.5 Decommissioning and Decontamination

The model requires that the estimated D&D cost be available at end of plant life. It is assumed that the D&D fund held in escrow by the seller (incumbent utility) is transferred to the buyer, who continues to hold it in escrow, and who picks up payments to a sinking fund to make the total equal the estimated D&D cost when the plant is to shut down.

The Excel future value algorithm is used to compute annual payments by the buyer. For the cases examined, these annual payments are trivial compared to other cash in and cash out flows.

Remarkably, much expensive Washington lawyer time and attention has been focused on the tax treatment of the change of hands of the escrow.

3.6 Operational Costs

The two operational costs include a) fuel and b) operating/maintenance (O&M). For fuel, the NPPACQ model uses lumped dollars per kilogram heavy metal as input, and assigns a dollar per year cost, based on the history of the plant in question. In detail, this is not correct; since nuclear fuel involves sizeable pre-use cash outflows for uranium, conversion, enrichment, and fabrication and transportation, these timing details of cash out flows should and will be included in future revisions of the model. A simple spot check for error introduced by the simplifying assumption, showed that IRR changed less than 5% (from 125% IRR), and annual cash flows to equity were less than 1% for a typical year. O&M costs vary significantly among plants. In most cases developed, we assumed a maximum of $85 million per year, which is a worse case scenario.

4. CASE I: REFERENCE CASE – A SECURITIZED TRANSACTION

For purposes of illustrating the NPPACQ model, a 1000 Mwe plant which came on line in 1982 with a capital cost of $1.5 Billion, is assumed. Table 1 provides the acquisition price calculation for the plant.

TABLE 1. Case I. Acquisition Price

(A) Original Capital Cost		***$***	***1,500.00***
(B) Plant Life (OL term)			***40***
(C) Annual Depreciation = (A)/(B)		$	37.50
(D) Commercial Operating Date (OL)			***1982***
(E) Years of Operation @	***2000***		18
(F) Undepreciated Value (Book Value)=(A)x[1-(E)x(C)]		$	825.00
(G) Securitization, if any		***$***	***625.00***
(H) Acquisition Price=(F)-(G)		$	200.00
Acquisition/BV			0.24

Note that the book value is above market value, which is assumed to be $200 million, the value at which the plant could compete, say, with natural gas fired generation. Thus the plant is a "stranded asset" and securitization of $625 Million is required. In a real case, market value might be determined by auction monitored by the state PUC, and resulting securitization requirement negotiated between the incumbent utility, the state PUC, and the buyer.

It is assumed that two annual fuel reloads in inventory, or about $60 Million, is included in the acquisition price. Thus the total acquisition price is $260 Million.

The plant had a seven-year construction period beginning in 1975, when a construction permit (CP) was granted by the Nuclear Regulatory Commission (NRC). When the plant entered service in 1982, the NRC granted a forty (40) operating license (OL), measured from start of construction. Thus the plant, assumed acquired in the year 2000 by the new owner, has fifteen (15) years of remaining licensed life. The plant license can be extended by the amount of the construction period (7 years) through a simple license amendment application to the NRC; to extend the license beyond this requires a more elaborate license extension process. For the example case, we assume that the plant will leave service after the nominal 15 years of license life remaining.

Table 2 gives other assumptions used in the example.

TABLE 2. Reference Case Assumptions

Capacity Factor	85%
Electricity Price	
Years 1-5	3.5 c/kwh
Years 6-10	4.0
Years 11-15	4.5
Plant Acquisition Capitalization (Plant and Fuel Inventory)	
Total	$260 Million
Debt Fraction	60%
Debt Term	15 years
Interest Rate	8%
Routine Annual CapEx	
Amount	$30 Million
Debt Fraction	75%
Interest Rate	8%
Major CapEx (Steam Generator Replacement, year 6)	
Amount	$130 Million
Debt Fraction	80%
Interest Rate	7%
Income Tax Rates	
Federal	34%
State	10%
Property Tax (Annual)	$8 Million
Operating Expenses	
Fuel	$30 Million/Year
O&M	$80 Million/Year
G&A	$10 Million/Year
DOE Waste Fee	1 mill/kwh
D&D Sinking Fund	
Total Estimated Requirement (15 years hence)	$500 Million
Current Year Balance	$100 Million
SF Interest Rate	7 %

Given the above assumptions, cash flow to equity (equity return) is presented in Table 3.

TABLE 3. Case I After Tax Equity Return

YEAR	REVENUE	OPERATING COST	INCOME TAX	CAPEX (EQUITY)	DEBT SERVICE	EQUITY RETURN	CUMMUL. EQTY RET
0	$0.00	$0.00	$0.00	$0.00	$0.00	($104.00)	($104.00)
1	$260.61	($144.36)	($36.30)	($7.50)	($20.95)	$51.50	($52.50)
2	$260.61	($144.36)	($34.72)	($7.50)	($23.80)	$50.22	($2.28)
3	$260.61	($144.36)	($33.12)	($7.50)	($26.79)	$48.84	$46.56
4	$260.61	($144.36)	($31.49)	($7.50)	($29.94)	$47.32	$93.88
5	$260.61	($144.36)	($29.80)	($7.50)	($33.29)	$45.65	$139.53
6	$297.84	($144.36)	($44.45)	($33.50)	($36.89)	$38.63	$178.16
7	$297.84	($144.36)	($32.28)	($7.50)	($58.23)	$55.47	$233.63
8	$297.84	($144.36)	($30.66)	($7.50)	($62.55)	$52.77	$286.40
9	$297.84	($144.36)	($28.91)	($7.50)	($67.41)	$49.66	$336.05
10	$297.84	($144.36)	($26.91)	($7.50)	($73.05)	$46.02	$382.07
11	$335.07	($144.36)	($40.88)	($7.50)	($79.84)	$62.48	$444.55
12	$335.07	($144.36)	($37.68)	($7.50)	($88.57)	$56.95	$501.51
13	$335.07	($144.36)	($32.68)	($7.50)	($101.19)	$49.34	$550.85
14	$335.07	($144.36)	($21.64)	($7.50)	($125.49)	$36.08	$586.92
15	$335.07	($144.36)	($75.69)	($7.50)	($18.23)	$89.29	$676.21
					IRR =	**48%**	

Note that the effective equity return (IRR) is 48%, after tax. Return on sales (ROS) is typically about 20%. Payback period is about two (2) years.

Figure 1 displays the Reference Case results graphically.

FIGURE 1. REFERENCE CASE EQUITY RETURN

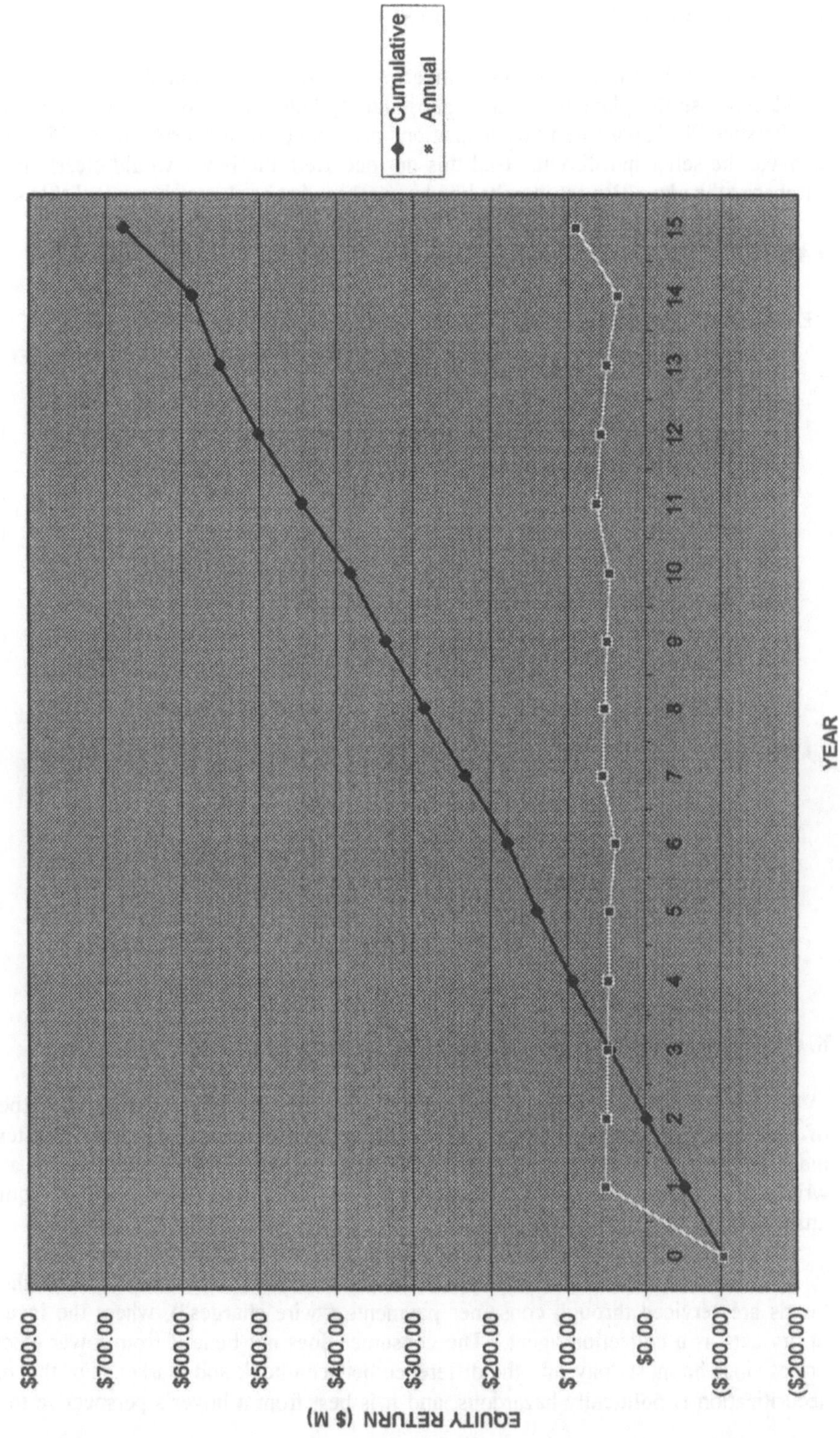

5. CASE II: REFERENCE CASE W/O SECURITIZATION

Case I resulted in a very robust investment performance, mainly because the buyer could purchase the plant for a value substantially below its book value ($825 Million), i.e. the state PUC sponsored "securitization" program provided a subsidy of $625 Million to make the seller indifferent. Had this not occurred, the buyer would clearly not have purchased the plant. Buyer's results had he purchased at book are shown in Table 4.

TABLE 4. Case II After Tax Equity Return

YEAR	REVENUE	OPERATING COST	INCOME TAX	CAPEX (EQUITY)	DEBT SERVICE	EQUITY RETURN	CUMMUL. EQTY RET
0	$0.00	$0.00	$0.00	$0.00	$0.00	($354.00)	($354.00)
1	$260.61	($144.36)	($4.76)	($7.50)	($64.77)	$39.22	($314.78)
2	$260.61	($144.36)	($3.68)	($7.50)	($67.61)	$37.46	($277.32)
3	$260.61	($144.36)	($2.60)	($7.50)	($70.60)	$35.55	($241.78)
4	$260.61	($144.36)	($1.53)	($7.50)	($73.75)	$33.47	($208.31)
5	$260.61	($144.36)	($0.46)	($7.50)	($77.10)	$31.18	($177.13)
6	$297.84	($144.36)	($15.77)	($33.50)	($80.70)	$23.50	($153.62)
7	$297.84	($144.36)	($4.31)	($7.50)	($102.04)	$39.63	($114.00)
8	$297.84	($144.36)	($3.47)	($7.50)	($106.36)	$36.15	($77.85)
9	$297.84	($144.36)	($2.54)	($7.50)	($111.23)	$32.21	($45.64)
10	$297.84	($144.36)	($1.45)	($7.50)	($116.86)	$27.67	($17.97)
11	$335.07	($144.36)	($16.39)	($7.50)	($123.65)	$43.16	$25.19
12	$335.07	($144.36)	($14.24)	($7.50)	($132.38)	$36.58	$61.77
13	$335.07	($144.36)	($10.37)	($7.50)	($145.00)	$27.84	$89.61
14	$335.07	($144.36)	($0.56)	($7.50)	($169.30)	$13.35	$102.96
15	$335.07	($144.36)	($55.93)	($7.50)	($62.04)	$65.24	$168.20
					IRR =	**5%**	

6. CASE III: A "MIDDLE-AGED LADY"

Cases I and II illustrated the importance, from both buyer's and seller's perspective, of state PUC sponsored securitization for stranded NPP assets. Securitization tends to make the seller indifferent to selling below book, in that it nulls the prospect of a major write-off. For the buyer, a purchase at or near market value makes the acquisition attractive.

Securitization, however, does not serve the electricity consumer because the state bonds are serviced through consumer payments ("wire charges"), where the incumbent utility acts as a collection agent. The consumer does not benefit from lower electricity prices since he must "pay off" the difference between book and market. For this reason, securitization is politically hazardous, and it is best from a buyer's perspective to target plants that do not require it, that is, plants whose book value is already at or below market value.

A class of US NPP's that are attractive acquisition targets are plants that came on line before the massive capital cost excursion of the late 1970s.[3] We label these plants "middle aged ladies". They came on line in the early and mid 1970s, many with modest initial capital costs. While they have less license life remaining than the later more expensive plants, their book values now, as depreciated, make them attractive acquisition targets. There are roughly 52 US plants that fit this profile.

The case presented below is a 1386 Mwe plant which came on line in 1973 with an original capital cost of $216 Million. Its current book value is $70 Million, well below market. Other assumptions are generally the same as in the Reference Case. Results are given in Table 5.

TABLE 5. Case III After Tax Equity Return

YEAR	REVENUE	OPERATING COST	INCOME TAX	CAPEX (EQUITY)	DEBT SERVICE	EQUITY RETURN	CUMMUL. RETURN
0	$0.00	$0.00	$0.00	$0.00	$0.00	($38.34)	($38.34)
1	$361.21	($162.80)	($76.12)	($20.00)	($17.08)	$85.21	$46.87
2	$361.21	($162.80)	($73.57)	($20.00)	($20.87)	$83.97	$130.83
3	$361.21	($162.80)	($70.94)	($20.00)	($24.85)	$82.62	$213.45
4	$361.21	($162.80)	($68.21)	($20.00)	($29.06)	$81.14	$294.59
5	$361.21	($162.80)	($65.38)	($20.00)	($33.53)	$79.50	$374.09
6	$412.81	($162.80)	($85.12)	($46.00)	($38.33)	$80.56	$454.65
7	$412.81	($162.80)	($71.61)	($20.00)	($60.97)	$97.43	$552.08
8	$412.81	($162.80)	($68.54)	($20.00)	($66.73)	$94.74	$646.82
9	$412.81	($162.80)	($65.14)	($20.00)	($73.22)	$91.65	$738.47
10	$412.81	($162.80)	($61.26)	($20.00)	($80.73)	$88.02	$826.49
11	$464.41	($162.80)	($79.28)	($20.00)	($89.79)	$112.54	$939.03
12	$464.41	($162.80)	($73.15)	($20.00)	($101.43)	$107.03	$1,046.06
13	$464.41	($162.80)	($63.85)	($20.00)	($118.25)	$99.51	$1,145.57
14	$464.41	($162.80)	($44.28)	($20.00)	($150.65)	$86.68	$1,232.25
15	$464.41	($162.80)	($127.77)	($20.00)	($13.44)	$140.40	$1,372.65
					IRR=	221 %	

[3] US nuclear plants under construction in the late 1970s suffered major capital cost escalation due to prolonged construction schedules, nuclear micro-regulation, high interest and escalation rates.

Return to Equity (IRR) after tax is 221%, and the plant earns over one billion in its remaining license life. Payback period on equity investment is less than one year.

7. GENERAL OBSERVATION: US NPP DEMOGRAPHICS and EQUITY CAPITAL FORMATION POTENTIAL

The brief analysis presented above raises several issues.

1. How many of the US NPP population offer capital generation potential similar to Case III above? I have examined 20 Case III type plants using data available from US Energy Information Administration (EIA) and other publicly available sources such as utility annual reports. There appear to be another roughly 32 plants in this category. The criteria was generally that book value (as best determined) was equal to or less that $300 Million for 1000 Mwe capacity.

2. What does this mean in terms of forward capital generation potential over the next 15 years? Simplistically, if 52 plants can deliver returns of about $1 Billion each over the next 15 years, then the total equity capital generation is $52 Billion. This is enough to launch a new wave of orders for standardized, pre-licensed NPPs, to complete the nuclear fuel cycle, as well as pick up where we left off on the US Breeder Program, assuming that this capital is in the hands of the private sector.

3. If equity earnings potential is so enormous for the Case III type plants, then why cannot the current utility structure achieve the same results, and why is de-regulation of power generation important?

    - The first and obvious answer is that in the current regulated regime, return on capital, even if potentially large, does not inure to equity holders' benefit, because ceilings are placed on investment returns in a regulated monopoly. As a corollary to this point, in the current regime, there is no incentive to maximize equity returns.

    - Second, entities dedicated to one technology, with equity return

    - As a driving force, operating in a competitive environment, are more apt to generated high equity returns through efficiencies, transparency, and fidelity to fewer complex objectives. Economies of scale applied to basic disciplines, and standardization of processes and functions will result in higher returns.

4. Why would incumbent utilities sell Case III assets if they are so profitable?

    0 Again, equity profit is not a driving force in the current structure.

    0 Several incumbent utilities (sellers) wish to divest nuclear generating assets to improve their bond ratings.

    0 In other cases, state utility restructuring legislation requires divestment, or at least unbundling and fire-walling the power generation function from other utility functions.

8. CONCLUSIONS

Nuclear industry attention should be riveted on the financial, legal, and institutional aspects of the US transition to a de-regulated power generation economy, rather than wistfully designing new reactors, longing for a new federal R&D program, and awaiting the second coming. While the de-regulation of power generation does not auger well for new construction in the near term, the new regime can be a capital generation engine to finance the nuclear renaissance. An "Adam Smith for the Atom" philosophy should be our new paradigm.

GLOBAL WARMING – AN OPPORTUNITY FOR NUCLEAR POWER? WELL YES, BUT.....

Peter Beck and Malcolm Grimston*

1. INTRODUCTION

A paper given by the authors at last year' Global Foundation Conference in Washington discussed the challenges to nuclear energy in this century and whether they can be overcome. The main conclusions were that they could be overcome, but with difficulties. Three aims were suggested:

- Reduce public suspicion of nuclear power expansion.
- Achieve a situation where the generation industry clamours to build new nuclear facilities.
- To make an impact on the global warming issue, nuclear generating capacity would have to expand considerably, say ten to twenty times today's scale by the second half of the century.

There have to be doubts whether these aims are achievable if only present technology were to be available. However, there are technological possibilities for developing plant and fuel cycles that would be more acceptable, but such development would require considerable R. & D. and commercial demonstration. As it is unlikely that the nuclear industry itself would be willing to fund such risky and expensive work, it may have to be done by governments, though in cooperation with the private sector, possibly even in multinational collaboration. However, it could well be that such funding would not be forthcoming unless public suspicion of nuclear energy is reduced. Hence the ' YES. BUT.....'of the title to this paper.

Since then, the RIIA started a two-phase project to study the future of civil nuclear power in greater detail. The first phase, now completed, identified those issues that may block the possibility of major worldwide expansion of nuclear energy and may even put at risk its survival. Five issues were identified as being of particular importance.

The Royal Institute of International Affairs, 10, St. James's Square, London SW1Y 4LE*

Global Warming and Energy Policy, Edited by Kursunoglu *et al.*
Kluwer Academic/Plenum Publishers, New York 2001

The second phase, only recently started, is to study these issues in greater depth. For that a series of questions are being circulated to a wide range of experts in the various fields, so ensuring that their contributions will cover a wide span of opinions and countries. Following discussions and possible workshops on the individual issues, papers on each will be developed. The ultimate aim of the research is to illuminate, not arbitrate, with the papers acting as a basis for briefing material for decision makers and opinion formers throughout the world.

This paper will review the results of Phase 1, which to a large extent confirmed the above conclusions, but also gave us a wider understanding of the main issues. The paper starts with a quick look at the overall energy scene and the possible roles of nuclear energy within that scene. It then continues by discussing each of the five main issues identified in Phase 1. Finally, the Conclusion shows the effect of this work on the conclusions of last year's paper.

2. THE ENERGY SCENE ONE YEAR ON

Although only a year has passed since the previous paper, there has been a subtle change in perception, largely brought about by the large increases in the price of oil from close to $10/bbl to close to $30/bbl. Whatever the reasons for these increases, they brought with them reminiscences of the 1970s, the fear of energy shortages, especially of oil and natural gas. Those arguing that oil and gas supplies may well start their rapid decline within the next few years, were once again listened to, even though others continued to maintain that there were ample resources of both and that the changes in price were largely due to the operation of market forces, acting on inadequate and often highly inaccurate information. The new perception brought with it renewed fears by countries of possible supply interruptions and hence concerns about energy security.

This change in perception may not be as drastic as it looks in N. America and Western Europe. A number of countries that are likely to become important energy users later this century, such as China, India, Korea, Indonesia, have never stopped their concerns about energy security. As with France and Japan, such countries tend to look at nuclear energy as a means of safeguarding their longer-term energy position. Russia, on the other hand, appears to have decided to make use of the possibility of oil and gas shortages by an expansion of nuclear energy for internal use, so freeing up oil and gas for export. Of course, there is no guarantee that these countries will carry out their plans; for example, shortage of capital may well interfere, but they do believe there is a strong logic to their aims.

Another change from last year is a stronger belief by many, though not all, countries that carbon dioxide emissions have an effect on the climate and should be controlled. Whether correct or not, the many droughts, floods and storms of the last twelve months are seen as the manifestations of global warming. This perception may be starting a shift towards more tolerance towards nuclear energy, at least to the point when the statement that 'we need to keep the option of an expansion of nuclear energy open' seems to be gaining ground, though still with the condition that this energy form has first to become acceptable to the public and profitable to the investors. It is of note that the recent IPCC energy scenarios include a number of cases where by mid century world nuclear capacity would be 10 to 30 times today's 360GW(e). Naturally, the requirement that nuclear power will have to become acceptable before such an expansion could become realistic is still made, but such views are is still a change from the previous position, when the IPCC appeared to ignore nuclear energy altogether.

Beyond these two developments, there have been few changes to the perception of the longer-term energy situation. Future demand seems as uncertain as ever and most scenarios still presume that demand will increase, especially in developing countries. There are still wide differences of views how far energy saving can keep the total energy demand in check. If greenhouse gas emission is to be restrained, far greater use of renewables, carbon dioxide sequestration and nuclear energy are seen as the means of reducing greenhouse gas emissions. There is, however, no resolution in the dispute between those who believe that renewables plus strong energy saving can on their own reduce carbon dioxide emission by the required amount and others who believe that all possible ways of producing energy without carbon emission will have to be used.

Should the climate change issue continue to climb up the political ladder of importance, it will be interesting to see whether the present pressures for deregulation and trusting the markets will continue. Can one really see the rapid development of renewables, which will, surely, require a great deal of highly risky demonstration projects in many countries, without considerable governmental support? And if there is such support, how will that affect the concept of deregulation?

Answers to such questions, which are likely to be in the political rather than strictly economic field, will be of considerable importance to the future of the nuclear energy. Will there be, as eventually there may have to be, a trade-off between the reduction of greenhouse gases and use of nuclear energy? Presently, the more extreme NGOs are able to ignore the obvious connection between the two and argue the need for rapid greenhouse gas reduction AND the immediate phase-out of nuclear power.

The hope that a more knowledgeable public and their representatives would be able to recognise such implausible logic is one of the driving forces for the RIIA's nuclear project. Another is the hope that more clarity about nuclear energy itself may stimulate similar work about renewables. That resource, which should be able to achieve a significant role within the future energy balance, is in danger of repeating the folly of nuclear power, which in the 1950s saw itself as the answer to all problems, only to fall from grace when it did not meet expectations.

3. THE RIIA "ENQUIRY INTO THE FUTURE OF CIVIL NUCLEAR ENERGY".

The purpose and a description of the project is as follows:

The noisy and acrimonious dispute about nuclear energy in many countries is such, that governments, industry and the financial sector find it more and more difficult to reach decisions in this field. Decisions require balanced and trustworthy information about the range of realistic options, at least for the main issues, and that is presently lacking. It is the intention of this project to provide such information from a standpoint of a neutral organisation with no interest in either camp, but close connections with both. It is not the intention to provide answers, such as what the future role of nuclear power might be, which, in any case, is likely to vary from country to country and should be left to decision-makers, not their advisors.

The project will seek to identify and clarify the major disputes in this field and suggest areas of possible resolution, where these may exist. The key output will be a programme of promulgating the findings to opinion formers and those who have to take decisions. Much detailed and valuable work has been and is being done in the area of nuclear power, environment and energy, but it is less clear that the results of such work have reached decision-makers. In effect, the aim is to identify what actions on the part of the nuclear industry, the financial sector and governments would allow different nuclear options to be kept open for different countries to meet future energy and environmental demands. For that, three scenarios have been developed: Slow exit, such as presently looked at by Germany, continuation of today's world-wide situation, assuming only small increases in capacity and major expansion by a factor of ten or more, as planned for in China

The project is organised in two parts. The first, which is completed, has developed a 'Position Paper', which identifies the main issues within the field of nuclear energy. The study is unique in taking into account the many disparate opinions both for and against nuclear powerfrom different organisations and a number of countries. This paper has been published in December 2000[α].

The results of the first stage have served to set priorities for Phase 2. That stage will involve the production of a series of research reports on these major issues. It is the intention to consider these central issues in depth rather than look at a large number superficially. The purpose will be to illuminate, rather than arbitrate.

4. LESSONS FROM PHASE 1.

The following issues were identified as those most connected with the future of nuclear energy:

- Public perception and the process of decision making
- Economics relative to other options
- Waste management, Reprocessing and Nuclear Proliferation
- Nuclear Research, Development & Demonstration (R., D. &D.).
- Safety

[α] Malcolm Grimston & Peter Beck: "*Civil Nuclear Energy - Fuel of the Future or Relic of the Past?*" Published by the Royal Institute of International Affairs, London, ISBN 1 86203 128 2, for Europe available from Plymbridge Distributors Ltd., Plymouth ,PL6 7PZ, UK and outside Europe from the Brookings Institution, Washington DC 20036 - 2188, USA.

There are, of course, a number of other important issues, but most of these impinge on one or more of the five. Initial lessons about these five issues, though to be carefully looked at and possibly modified in Phase 2, are as follows:

Public Perception and the process of decision-making: There is the perception within a number of OECD countries that there is considerable antipathy against nuclear power. The evidence for that is, however, not all that strong, but what matters are perceptions, whether true or not, and the effect these may have on a country's decision-makers. The table below, based on a MORI Poll in 1999 in UK, suggests that assumption and reality can be far apart and there are indications of similar differences in other countries. The reasons for such discrepancies may well be associated with the headline-catching activities of the opponents against the defensive stance of the industry. Breakdown of a nuclear facility is news; continuing good operation is not. In some countries opponents demonstrate, at times violently; protagonists do not and therefore do not grab the headlines.

PUBLIC OPINION ABOUT NUCLEAR ENERGY, MORI, UK, 1999.

	Favourable towards nuclear energy industry	Unfavourable towards nuclear energy industry	Neither favourable nor unfavourable/ don't know
Public opinion	28 %	25 %	47 %
All MPs	43 %	44 %	13 %
MPs' perception of national public opinion	2 %	84 %	14 %

A further problem affecting public perception is the fact that in most countries that have deregulated and privatised the electricity market, the electricity industry is not agitating for more construction of nuclear plants. They wish to keep existing plants in operation for as long as possible, but are uninterested in new plants. In countries where governments are committed, for whatever reason, to build more nuclear capacity, (India, China) public perception may well be quite different. Should pressure mount from the power industry in OECD countries that more nuclear power would provide benefits to the country and to their customers, the public's and decision makers' perception may well change.

An important measure of international perception about nuclear power will be the outcome of the present debate whether or not nuclear power would be accepted as meeting the criteria for the Clean Development Mechanism (CDM) of the Kyoto Protocol. Those criteria imply that CDM projects must be seen as sustainable development, whilst also achieving reductions in greenhouse gas emission. According to the nuclear industry, nuclear power should qualify, according to opponents it should not. There is little doubt that acceptance of nuclear power would greatly enhance its case and vice versa.

Economics: Two factors have, over the last ten years substantially harmed the economics of nuclear power compared to natural gas. The first is the development the combined cycle gas-turbine (CCGT) generating plant, with its higher efficiency, lower capital cost per unit and greater flexibility. The second is the liberalisation of the electricity markets in many OECD countries. In the past, power companies had an effective monopoly within a region, but they were strictly regulated. This meant that any additional costs, as long as accepted by the regulator, could be passed on to the customer. In a competitive structure this situation no longer exists. There is no guaranteed market and therefore no guaranteed off take from a new plant and any unexpected expenditure or loss of markets has an immediate effect on the bottom line.

Under such circumstances, a large-scale nuclear plant with high capital cost, taking perhaps eight years from concept to start-up, is likely to frighten off the investor. Where natural gas is available, the alternative could be a CCGT plant with far lower cost per unit and a construction period of perhaps two years. Unless governments decide to intervene for environmental and/or energy security reasons, new present types of large scale LWRs have in liberalised markets become very risky investments.

Should there be restriction on the use of fossil fuels, the main competition to nuclear power is likely to come from renewables, but there is little reliable information about the costs of the various renewables when utilised on a large scale, on which to base a worthwhile competitive analysis.

In summary, if nuclear power is to be certain of being able to compete, the industry may well have to set itself a target of matching CCGT plant economics, construction time and flexibility of scale, a target that is understood to be aimed at by the S. African ESKOM Pebble Bed Modular Reactor (PBMR) project

Waste Management, Reprocessing and Nuclear Proliferation: These issues have been combined because of their close interlinkage. Most statements against nuclear power mention that the industry has no answer about the ultimate destination of long-lived radioactive waste and that nuclear power should not be allowed to expand or even to continue unless the issue is resolved in a way acceptable to the public. The industry denies this lack of solution and maintains that deep underground repositories are the best answers, but that argument is fatally weakened by the fact that after over twenty years work in a number of countries, there is still no such repository in operation. Furthermore, NGOs argue that these repositories could become the plutonium mines of the future and that today's technology is in principle quite inadequate to ensure that such repositories will be secure for a hundred thousand years or so. The uncertainties about waste may well be one of the strongest weapons in the armoury of the opposition.

As no repository is likely to be ready for many years, intermediate storage facilities will ***have to*** be built in a number of countries. There are those who argue that such facilities should be designed to last at least one hundred years, so giving the industry time to find on optimal solution, which may well be quite different for different futures of nuclear energy and for different countries.

If the option of major expansion were to be kept, issues such as uranium availability, the pros and cons of reprocessing and the present idea that every country has to look after its own waste, implying a local repository, have to be carefully examined. Other means of dealing with the problems may well have to be found, such as regional repositories and/or transmutation. The effect of the various solutions on the dangers of proliferation also needs careful assessment.

One prerequisite for all such solutions may well be the need to be able to ship waste over country borders to repositories and/or transmutation plants. That is presently not allowed and there is no doubt that any change to that law will be bitterly resisted by dedicated opponents of nuclear power.

In summary, achievement of an acceptable solution to the waste problem may well be 'a must' if nuclear energy is to have a better chance.

Nuclear R. D. & D: One of the results of deregulation and privatisation was the decision by many Governments to cut energy R. & D. spending, possibly on the assumption that private corporations will increase their expenditure. Instead, private companied also cut, so that with the exception of Japan, total expenditure dropped heavily in all major developed countries. Reduction in nuclear R. & D. was especially high; thus, in USA 1997 expenditure on nuclear fission R. & D. was 3.7% that of 1978.

As matters stand today, there is continuing research on short-term schemes, such as improvement of fuel and lifetime extension of plant. There is some expenditure on longer-term projects, but, except possibly for the Asian countries, there is little enthusiasm for going ahead with demonstration plant, an essential step before commercialisation.

Yet, if nuclear energy is to expand significantly, work may be needed inter alia on:

- Smaller, safer and cheaper reactors, more suitable for the developing world and competitive with CCGTs as well as renewables,
- more proliferation resistant fuel cycles,
- methods of resolving the waste issue,
- the possibility of extracting uranium from seawater,
- new approaches to reprocessing.

Some of such R. & D., say on waste disposal, would be needed whatever the future of nuclear power.

Laboratories have developed many leads for such work, but the question is how further development towards commercialisation could be funded. Major international collaborations have been suggested to cover not just nuclear power, but all non-carbon energy sources, but so far there is no flesh on the bones of such ideas.

As matters stand, countries such as Japan, India and China may well continue major research, possibly in conjunction with funding work in Russia. These countries seem concerned that the West may not continue it's development role in the nuclear area. If they are to depend on nuclear energy to improve their energy security, they believe that they may have no choice but to forge ahead with R. & D. themselves. The implication of this could be that the leadership of technology and therefore of global policy on nuclear energy will slowly move from USA and (to a lesser degree) Europe to Asia.

Safety: There is no doubt that safety of nuclear installation and of transport of nuclear material is a key issue. The shadow of Chernobyl still lies over the industry and it is clear that another major nuclear incident would have substantial implication worldwide. The nuclear industry, unlike a number of other technologies, cannot afford to learn through major mistakes.

Safety depends not just on the quality of design, but perhaps to an even larger extent on the quality of management of operation, transport and maintenance. Before major expansion, including into developing countries, could be contemplated, acceptable safety will have to be ensured, a difficult task, as absence of accidents, so far, cannot be a guarantee for no accidents in future.

Present designs largely depend on 'engineered safety' but significant expansion of nuclear energy may well require 'passively safe' designs, where safety would not depend on either operator action or on the operation of mechanical or electric devices. Designs for such plant have been developed, but so far, no demonstration facilities have been built. There is a school of thought which argues that such plant could become far simpler than present reactor designs, so reducing capital cost and hence improve the economics of nuclear facilities.

5. CONCLUSIONS

Although the study, so far, has corroborated most of the points of the previous paper, some issues are now looked upon in a fresh light:

- **Energy Security** is coming to the fore again. The fact that the future energy situation is highly uncertain, although always the case, appears to be more accepted now and that has increased concerns about security. Use of nuclear energy can improve such security.
- **Public Antagonism** against nuclear energy may not be as solid as it appears to many politicians and decision makers. Opponents know how to make headlines, industry has far fewer opportunities - making headline news out of safe operation is difficult.
- **To be wanted** by the generating industry and governments may well be an important way to achieve greater acceptability. Opponents maintain that nuclear power is quite uneconomic and that may well be true in those countries with access to natural gas. The industry needs to set itself targets to be competitive with CCGTs and not solely depend on its advantages due to no carbon dioxide emissions. In any case, when it come to these advantages, the competitor becomes renewable energy, the economics of which, when used on a large scale, are still unknown.
- **Waste management** may well be the Achilles heel of the nuclear industry. The industry has little to show for 20 years work and opponents make the most of this. Large scale expansion of nuclear power may well be quite impracticable

unless sufficient feedstock becomes assured and if the concept continues that every country has to find its own solution for waste within its own borders. The issue of proliferation is another concern in this area.

- **Nuclear R. & D.** would have to be much strengthened if the option for major expansion were to be kept open. There are potential solutions for most of the concerns holding back nuclear energy, but their resolution will require much R. & D. and erection of demonstration plants. As the West seems under present circumstances unwilling to carry out such a programme, it may be left to Japan, China and India to do so, although it has to be seen whether they will be able to fund such a programme. As long as funds can be made available, Russia may also be able to play a significant part. Such a course would move the technological leadership of civil nuclear energy from USA to Asia

Phase 2 of the project will look at these conclusion in greater depth.

NEAR-TERM DEMONSTRATION OF BENIGN, SUSTAINABLE, NUCLEAR POWER

Carl E. Walter*

Nuclear power reactors have been studied, researched, developed, constructed, demonstrated, deployed, operated, reviewed, discussed, praised and maligned in the United States for over half a century. These activities now transcend our national borders and nuclear power reactors are in commercial use by many nations. Throughout the world, many have been built, some have been shut down, and new ones are coming on line. Almost one-fifth of the world's electricity in 1997 was produced from these reactors. Nuclear power is no longer an unknown new technology.

A large increase in world electricity demand is projected for the coming century. In lieu of endless research programs on "new" concepts, it is now time to proceed vigorously with widespread deployment of the best nuclear power option for which most parameters are already established. Here, we develop an aggressive approach for initiating the deployment of such a system— with the potential to produce over half of the world's electricity by mid-century, and to continue at that level for several centuries.

REACTOR FACTS AND PROJECTIONS

At late count[1] 434 reactors, each with an average capacity rating of over 800 MWe were in operation in 34 countries. An additional 62 reactors (~15%) are under construction or on order. These will have a slightly larger average capacity of 850 MWe. Also, 77 reactors are no longer in service and will be decommissioned. On the average, the latter are older, smaller capacity (~330 MWe) reactors. The trend is to build larger reactors and to close down the older, smaller reactors. Almost 80% of the world's reactors now in operation are light water reactors (LWRs) and almost 75% of those are pressurized. In the U.S. 100% of the power reactors are LWRs. Thus, it appears that LWRs are an established technology in the current global nuclear electric power infrastructure.

Nevertheless, there appears to be a small but dominant public perception that nuclear power is unsafe and can lead to the use of nuclear weapons in the future. In consideration of this public perception, those who project future consumption of electricity tend to limit their projections with respect to future nuclear electric capacity and consumption.

* Lawrence Livermore National Laboratory, P.O. Box 808, Livermore, CA 94551

Global Warming and Energy Policy, Edited by Kursunoglu *et al.*
Kluwer Academic/Plenum Publishers, New York 2001

As a result, there is considerable uncertainty on the future use of nuclear electric power in the U.S. as well as in some other countries. Sweden and Germany have decided to shut down all of their nuclear power reactors in response to political pressure from those opposed to nuclear technology.

The Energy Information Administration (EIA) projects, in its reference case, that U.S. nuclear power capacity in 2020 will be reduced to 58% of the 1997 value, although because of continuing improvements in operation, the electricity supplied will be down to only 68%. The corresponding projected reductions on a world basis are less severe, 86% and 94%. These nuclear power projections are made in the context of increased electricity consumption projections in the U.S. and the world of 33% and 76%, respectively. These data, based on EIA projections[2] are shown in Table 1. Thus, by the year 2020, the projections in net electricity consumption indicate an average annual increase since 1997 of 1.2% in the U.S. and 2.5% in the world.

Increased electricity use is projected throughout this century. Based on composite longer-range projections from various sources,[3] conservative estimates of electricity consumption during this century are shown in Table 2. An average annual growth rate of 1.4% is projected for world electricity consumption, thus quadrupling consumption during this century. U.S. electricity consumption is projected to increase at a considerably lower rate, resulting in about 70% higher consumption by 2100.

THE NEXT STEP

With the seemingly ubiquitous and productive LWR technology in place throughout the world, one might assume that only minor advancements in reactor design could be expected. Although design changes appear to be minimal, substantial improvements in advanced LWRs result in even safer and more economical systems than the current fleet. Despite their many desirable features however, advanced LWRs do not provide a sustainable technology. Some advanced LWRs are being built in Japan and in Korea

Table 1. Current and near-term projected nuclear power capacity and consumption and total electric power consumption for the U.S. and the world.[2]

	1997	2020	ratio[a]
U.S. nuclear generating capacity, GW	99.0	57.0	.58
World nuclear generating capacity, GW	351.9	303.3	.86
U.S. nuclear power consumption, GW	71.8	48.7	.68
World nuclear power consumption, GW	258.9	243.8	.94
U.S. nuclear plant capacity factor, %	72.5	85.4	
World nuclear plant capacity factor, %	73.6	80.4	
U.S. net electricity consumption rate, GW	374	497	1.33
World net electricity consumption rate, GW	1400	2463	1.76
U.S. nuclear share of electric consumption, %	19.2	9.8	
World nuclear share of electric consumption, %	18.5	9.9	

[a] ratio of value in 2020 to value in 1997

Table 2. Century projections[3] of annual net electricity consumption rate.

	2000	2025	2050	2075	2100
U.S. net electricity consumption rate, GW	380	480	560	610	640
World net electricity consumption rate, GW	1540	2530	3570	4640	6140

where public opposition to nuclear technology has not resulted in extended, very costly, construction duration, as was the case in the U.S. with the current fleet of LWRs. Improved licensing procedures for generic advanced LWRs are now available in the U.S. These procedures are meant to eliminate licensing delays and the attendant high construction costs that have occurred previously. Unfortunately, skittish investors have not initiated plans for construction of a nuclear power plant in the U.S. for a quarter century— the new U.S. licensing procedures remain untried.

Without pursuing a program to facilitate the deployment of advanced LWRs in the U.S., the Department of Energy (DOE) has dubbed[4] them Generation III and seemingly dismissed them as contenders in the U.S. DOE considers the advanced LWRs to be insufficiently cost-competitive in a deregulated electricity market, faced with unresolved used fuel disposition plans, and a potential means of nuclear weapon proliferation.

As a result, DOE has launched its Generation IV program,[4,5] to consider reactor designs and fuel cycles that 1) are even more resistant to nuclear weapon proliferation than the once-through cycle used with LWRs, 2) minimize radioactive material waste and utilize publicly accepted and implemented waste solutions, 3) provide electricity competitively priced with other forms of generation with acceptable risk to capital and having short lead and construction times for new plants, 4) have low likelihood of core damage and no severe damage for plausible initiating events, and 5) meet specified safety criteria. The Generation IV program[5] envisions research and development (R&D) on various reactor types such that a prototype plant could be operated by 2020, and a large scale deployment by 2030.

It is the thesis of this paper that the exploratory R&D on various types of new reactor concepts is unnecessary, and in fact incompatible with the large deployment schedule objective. Instead, a program to demonstrate the already highly developed modular fast reactor with fuel recycling should be vigorously pursued. Such a technology is sustainable for centuries. Moreover, it is ethically correct from the standpoint of energy resource conservation and stewardship of residue waste. It can meet all of DOE's Generation IV objectives and would allow beating DOE's target date for large-scale deployment by at least ten years. That should be "The Next Step."

Before examining the status of fast-reactor/fuel-recycling technology and means of initiating its deployment on a large scale, we address the environmental and resource considerations leading to its choice, as well as some issues that are the basis for negative public perception of nuclear power. Public perception needs to be corrected throughout the world to allow global progress on implementation of any nuclear power option.

ENVIRONMENTAL AND RESOURCE CONSIDERATIONS

Fast reactors with fuel recycling do not require uranium enrichment. Natural uranium would not need to be mined for a very long time. Instead, depleted uranium (two million tons are projected[6] to be stockpiled at enrichment plants by 2015) would be used as makeup material for new fuel elements. The number of recycles is unlimited, therefore discharged fuel is never in need of permanent storage. The waste product from the fuel recycling process can be designed to contain no significant amounts of actinides or long-lived fission products.

Geologic disposal would continue to be desirable for the non-recyclable waste from the fuel recycling process. However, the disposal facility requirements for this waste form would change significantly from those now being considered for used fuel from LWRs. Essentially all the long-lived fission products could be selectively removed in the recycling process (and subsequently transmuted). Waste from fuel recycling would contain no actinides, so there would not be a requirement for long-term safeguards against material diversion. Thus, the time horizon for the geologic disposal site would decrease many thousand to ~500 years. Such disposal is respectful of future generations.

Systematic retirement of LWRs as they complete their design life and their replacement with fast reactors is an appropriate evolutionary advancement in electric power generation. All of the LWR used fuel would be processed and utilized in new fast reactors. This is the technology that needs to be demonstrated as quickly as possible through construction and operation of a prototype reactor producing power to the grid.

Nuclear power reactors of any type generate electricity without the carbon dioxide emissions released to the atmosphere from fossil fuel power plants. In 1997, carbon emissions from electricity generated[7] in the U.S. contained 532 Tg C. Had not almost 20% of the electricity been generated by nuclear plants, another 151 Tg C (~28%) would have been emitted by the additional fossil power plants that would have been required. In view of the concerns about climate change, and the possible contribution to this effect by carbon dioxide in our atmosphere, nuclear power generation has a significant positive effect on the environment.

FISSILE MATERIAL DIVERSION ISSUE

Both the nuclear weapon and nuclear power communities appear to use the single word "proliferation" or the words "nuclear proliferation" with the implicit understanding that it is in the context of nuclear *weapon* proliferation and therefore, something to be avoided. On the other hand, *proliferation* of nuclear power reactors seems to be just what is needed to solve our present and expected future electricity shortages without harming the environment. The word *proliferation* can signify a good thing! How is the general public to know what is meant if *proliferation* is used without the intended modifier? The message that the public appears to get is that proliferation of *all nuclear technologies* is a bad thing.

In any case, the public's concern should not be proliferation of nuclear weapon capability to additional countries. The public's concern should be the potential for diversion of fissile material from its intended use in a power reactor fuel cycle for making nuclear weapons or crude nuclear explosives by individuals that act with or without the approval of the material owner. Thus, it is imperative that fissile materials can nowhere be diverted overtly or clandestinely for the manufacture of nuclear explosives. This can be accomplished with intrinsic physical/chemical characteristics of the technology that is used, together with oversight by international review organizations, such as the International Atomic Energy Agency. The public must not be swayed into a paranoid position precluding peaceful uses of fissile materials and nuclear technology. Nuclear technology offers too many advantages to the health and general well being of humankind and our world environment to ignore.

Fast reactors with fuel recycling can provide electricity for centuries at a competitive cost in a manner respectful of the environment. Fuel recycling would utilize relatively inexpensive pyro-metallurgical processing and solvent electro-refining. The considerably more expensive aqueous processing previously proposed or used in the nuclear weapon and nuclear power communities was developed to produce pure plutonium. On the other hand, pyro-processing is inherently resistant to material diversion. At no time does pure, separated, plutonium exist. The presence of minor actinides in the plutonium makes it unusable directly for a nuclear explosive, as the actinides produce heat and radiation and preclude, or greatly impede, the construction of an explosive device. Some fission products remain with the plutonium, preventing hands-on theft. Process waste would not contain significant amounts of actinides and therefore, is not at all an attractive material for making nuclear weapons.

USED FUEL/WASTE ISSUE

The media and even those in the nuclear field use unclear terms relative to various aspects of nuclear technology. Use of inaccurate or ambiguous terms complicates the achievement of public acceptance of nuclear matters. For example, the term "spent nuclear fuel" uses the word *nuclear* gratuitously, unnecessarily inciting fear in the public. There is no other kind of "spent" fuel making the modifier *nuclear* necessary. Also, the terms "spent fuel" and "nuclear waste" are often incorrectly used synonymously—although the used fuel from LWRs is barely "spent" in an energy sense and should not be considered to be waste. In the past, less than one percent of the energy potential of mined uranium has been utilized in LWRs.

By whatever name we call used fuel, nuclear technology has advanced to the point that, with the appropriate reactor and fuel design, there should be no used fuel to dispose of in a geologic repository. Only a small amount of radioactive waste with insignificant amounts of actinides and a much reduced half life resulting from the advanced recycling process employed would need to be disposed of in a geologic repository.

RADIATION HEALTH EFFECTS ISSUE

A critical issue that needs to be resolved in the minds of the public concerns radiation health effects. First, a preponderance of scientific experts in the field must come to agreement. At present there is disagreement among experts regarding the effects of radiation at low levels, below ~50 mSv/y. One group believes that radiation effects are linear and that there is no threshold below which radiation is harmless. A larger group believes that there is a threshold level below which radiation is harmless, and some in this group believe that low-level radiation is in fact beneficial. Clarification of this issue is essential so that unambiguous information can be presented to the public. In view of the scientific discord, the matter has been politicized and even U.S. agencies disagree among themselves.

Resolution in favor of an acceptable threshold will have a positive effect on public attitude regarding nuclear power. Also, the acceptable radiation level that is promulgated must be based on a risk/benefit analysis in the context of other anthropogenic sources. Currently, radiation standards are established far below highly variable natural radiation levels that in the U.S. average 3 mSv/y, and without reference to higher risks from other sources that society now accepts. The unresolved controversy was recently the subject of a review by the General Accounting Office.[8] At present the National Academy of Sciences, through its committee known as BEIR VII (Biological Effects of Ionizing Radiation- No. 7) is reviewing the matter and plans to issue a report next year, but it is doubtful that that schedule will hold. In addition to providing the public with an accepted scientific basis for a radiation standard, considerable savings (reducing the cost of nuclear power) can be realized if the standard is not unnecessarily low.

THE FAST REACTOR SOLUTION

Fast reactors have been operated successfully at DOE installations for a number of years. Originally, it was thought that fast reactors, while also generating electric power, would be used to produce excess plutonium. The excess pure plutonium and uranium from LWR used fuel would be mixed to provide the necessary fissile content in new mixed oxide (MOX) fuel for LWRs. The nuclear power infrastructure would consist of both fast and thermal reactors and fuel recycling facilities, thus necessitating public transportation of new and used fuel. Exclusive use of fast reactors with on-site fuel recycling, that precludes production of pure plutonium now appears to be the better option.

The Advanced Liquid Metal Reactor (ALMR) program in the U.S. was in full swing and making good progress until 1994. The program goal was to develop a modular fast reactor, sodium cooled, and utilizing metallic fuel elements. The fuel would be recycled at the reactor site using improved pyro-metallurgical and electro-refining processes. Process waste would contain insignificant amounts of actinides and would not be an attractive material diversion target. This program was identified in the 1992 Energy Policy Act as dealing with a key nuclear technology that should be supported with R&D funding for a five-year period to enable future decisions regarding its course.

Unfortunately, under the negative view of the new Administration beginning in 1993, Congress did not appropriate the necessary funding to support the ALMR program after 1994 and in fact ordered the DOE to terminate the program in February 1994. On February 17, 1993, President Clinton had stated in his first speech[9] to a joint session of Congress that: "We are eliminating programs that are no longer needed such as nuclear power research and development." The next day, at a public address[9] in St. Louis, he expanded on his previous evening's talk: "We recommended some unwarranted subsidies be eliminated because the need for the work is much less or nonexistent anymore. For example, we recommended a big cutback in a lot of programs related to the nuclear industry and the elimination of a nuclear research program that is inconsistent with our new energy future." As a result, DOE's ALMR program was cancelled. Work at the Argonne National Laboratory (ANL) on the fuel cycle was suspended, as well as DOE-supported industrial design studies on a commercial power reactor/fuel recycling system by General Electric Co. (GE) and others, such as Burns and Roe.

Fortunately, GE continued its design studies with company funding and with support from Tokyo Electric Power Co. The current reactor design is called Super PRISM. ANL has been able to corroborate some of the fuel processing parameters that make fuel recycling viable. This information was gained as a result of performing some necessary fuel treatment resulting from the directive to shut down the experimental fast reactor in Idaho. As a result, there are sufficient data to proceed with the construction of a demonstration power reactor plant.

The Super PRISM design[10–12] embodies a number of features that appear to resolve many issues of concern. Particular attention has been given to reactor safety. Passive features minimize the need for operator action and expensive backup cooling systems. Considerable savings in the cost of technology demonstration and system hardware are realized because of the modular design. A plant is composed of one to three power blocks, and each power block couples two 1000-MWth fast reactors coupled to a single-superheat turbine/generator system producing a net electrical output of 760 MWe. Thus a full size plant (six modular reactors) would produce a net output of 2280 MWe at 38% thermal efficiency. Each module has the ability to operate independently of others. Module size was selected on the basis of constraints on factory fabrication of the reactor/containment vessel.

Although as noted above the trend has been toward higher power LWRs, the modular scaling of Super PRISM is actually cost advantageous. The modular design avoids much field construction effort (reduced field time) in view of extensive factory fabrication, and allows the design to be simplified through the use of passive shutdown heat removal and passive post-accident containment cooling systems. Demonstration of the smaller, but prototypical reactor can be accomplished at reduced cost. The low cost of power also results from the higher capacity factor achieved because of modularity; generic licensing regulations; elimination of active safety systems because enhanced safety is provided by passive features; and simplicity that results in lower operation and maintenance requirements. These features all contribute to the lower cost of electricity.

Super PRISM could be operated with metal or oxide fuel and at a variety of conversion ratios. The fuel cycle using metal fuel and a low conversion ratio minimizes costs. Therefore the demonstration reactor would most likely have metal fuel and operate at a conversion ratio slightly above one (breakeven).

Specifically, it is proposed to demonstrate the operation of a Super PRISM module utilizing its considerable database. A recent economic analysis[11] of the Super PRISM design indicates that the cost of power, \$28/MWh for the nth of a kind plant, is easily competitive with other types of power. There doesn't appear to be a good reason to conduct exploratory R&D as proposed in DOE's Generation IV program, referred to earlier, instead of proceeding at once with demonstration of Super PRISM. Some confirmatory R&D in support of the Super PRISM design would be warranted, however.

NEAR-TERM REACTOR DEMONSTRATION

At a recent conference, Daniel Fessler,[13] former Chairman of the California Public Utilities Commission, alluded to an imminent shortage of electric power in Brazil at the end of year 2000 as an example of the worldwide need for clean responsible sources of electricity. He challenged the conference participants to move nuclear power construction out of its apparent hiatus and thus rise to the challenge of the projected worldwide electricity needs. He emphasized the urgency for the nuclear community to proceed on these matters with the comment that "...one shouldn't wait ten weeks to begin a task that must be completed within ten years."

Mr. Fessler's remarks give rise to serious thought and a resulting multifaceted solution to advance the acceptability of nuclear power technology throughout the world. Demonstration of Super PRISM in any country, under international auspices, would be of benefit to all countries. Such a project could be accomplished in Brazil, or in another suitable country, within a ten- to fifteen-year time frame.

THE BRAZIL DEMONSTRATION VENUE

Although the fast reactor/fuel recycling demonstration project could technically be conducted anywhere in the world, it would be desirable to perform the demonstration in a developing country, preferably, from a neighborly viewpoint, in the Western Hemisphere. Countries, such as the U.S., France, and Japan, where nuclear power is in relatively wide use, have run into public perception problems regarding new nuclear projects. Although perceptions are changing, the public in these countries does not currently appreciate the need to develop advanced nuclear power plants.

By various measures Brazil is a key world country. It ranks fifth in land area and in population among the world's ~220 countries. Brazil is a significant generator and consumer of electricity. In 1998, Brazil generated 36.2 GWy of electricity, ranking tenth in electricity generation among world countries. The top ten countries together generated over two-thirds of the world electricity. Over 90% of Brazil's electricity is hydroelectric. Brazil's per-capita usage will more than double by 2020 according to U.S. projections.[2] Brazil's population is projected to increase less than 25% by 2020. In the same projections, however, electricity consumption is predicted to increase by a factor of 2.8, carbon emissions by 2.6, gross domestic product by 2.4, and total primary energy use by 2.1.

Until recently, Brazil had only one operating reactor, Angra 1, a 657-MW Westinghouse PWR. Lifetime performance of Angra 1 since its startup in 1982 has suffered. However, its lifetime load factor of only 28.1% improved to 60.9% for the year ending June 1999. A second reactor, Angra 2, a 1300-MW Siemens-KWU PWR, was recently completed. It achieved first criticality on July 14, 2000 and 100% power soon after. Work had been suspended on a third reactor, Angra 3, similar to Angra 2 at the same site. A financial strategy is being developed for continuation of the Angra 3 project that would commence operation in 2006. These two larger units would add ~6% to the 36.2 GW of electricity generated in Brazil in 1998. Longer-term plans through 2015 include a possible Angra 4 reactor (similar to Angra 2 and 3) and a fifth reactor of an unspecified type. A Super PRISM demonstration project in Brazil is not inconsistent with continuation of those plans, as used fuel from the LWRs could be applied to startup inventories of fast reactors that would most likely be adopted in the future.

Because currently Brazil's electricity is mostly hydroelectric, carbon emissions from electricity generation are practically negligible. In 1990, carbon emissions from electricity generation in Brazil were 0.09 kgC/Wy compared with 1.54 kgC/Wy in the U.S. Clearly, Brazil's electricity has not contributed to climate change. But dams are no longer popular in Brazil. Agricultural interests are said to be harmed by further dam construction; rainforests would be flooded thus impacting the world environment; and long distance electricity transmission would be required because of the remote location of the new dams that would be constructed.

Brazil has limited fossil fuel resources. Oil provides primary energy at the rate of 85 GW, with almost 40% of that imported. Indigenous coal and gas contribute only ~20% as much primary energy, and ethanol, made from sugar cane, contributes another ~20%. Recently a 3000-km natural gas pipeline, costing $2B, was built from Bolivia to Brazil. Its current capacity is about 11 billion m^3/y or about 13 GW of primary energy. If used exclusively for electricity generation in highly efficient combined-cycle gas plants, 7 GW of electricity could be generated, or less than 20% of the 1998 generation. It should not be assumed, however, that the cost of electricity from such gas plants would be lower than from a Super PRISM plant.

If the admirable carbon emission performance of Brazil is to continue, new electricity capacity from nuclear and renewable power resources will need to be brought on line to meet the ever-increasing requirement for electric power. A clear solution to Brazil's electricity needs appears to be expanded use of nuclear power. Nuclear power is the leading contender as a future substitute for hydroelectric power, if carbon emissions are to remain low. In view of Brazil's enviable record on carbon emissions from electricity generation, and the projected competitive cost of power through the use of Super PRISM technology there does not appear to be a better solution.

Although Brazil's experience with nuclear technology is limited, they are a signatory of the Nuclear Weapon Non-Proliferation Treaty, have considered development of nuclear powered submarines, have developed isotope enrichment technology, and accept international regulations for safeguards of nuclear material. The initial inventory of fuel for the Super PRISM demonstration reactor could be provided from the used fuel projected to be discharged from Brazil's LWRs by 2010.

Because of Brazil's historically limited nuclear power exposure, demonstration of a fast reactor/ fuel recycle system in Brazil should meet with less public opinion resistance than in countries with a more entrenched anti-nuclear technology sector. Brazil's electricity needs would be met, and also the world public would benefit from the Super PRISM demonstration and its subsequent deployment throughout the world.

PROJECT INITIATION AND MANAGEMENT

In order to initiate the demonstration, some funding is required to conduct preliminary studies of specific issues, establish the participants, develop a project plan, and secure approvals and funding. A key goal of this initial effort would be to win the genuine interest of the host government (for example, Brazil) in the demonstration project and obtain its permission and facilitating support for conducting the project. The resulting project plan would form the basis for a "full-ahead" direction to proceed. Success of the Super PRISM demonstration is highly dependent on the quality of its leadership. The participants would clearly include a number of industrial and government organizations from throughout the world. Active participation by university personnel and of nuclear technology associations would be encouraged.

The project should be conducted as much as possible on a private business basis in order to avoid government entanglements. Experience gained from this project would benefit various world governments, businesses, and organizations. Because of first-of-a-kind risks and its global importance, their subsidies should be solicited to defray part of the cost of the Super PRISM demonstration.

IN CONCLUSION

Although the suggested demonstration project of Super PRISM cannot fill short-term needs for electricity, an immediate start appears essential. This urgency is dictated by the continuously increasing use of electricity, throughout the world and the global need to maintain low carbon emissions. Every effort should be made to complete the remaining development of the Super PRISM reactor and complete construction of the demonstration reactor by 2015 at the latest to allow wide-scale deployment to be in place by 2030, consistent with DOE's Generation IV objectives. With an early project start, a qualified management, and dedicated world project team players, such a schedule, needed to satisfy future world electricity demands could be met. In fact, it appears that Super PRISM is the only viable technology that can meet the requirements recently set forth by the nine-nation group that includes U.S., Brazil, and Argentina for Generation IV reactors.

The ultimate desired outcome of the Super PRISM demonstration project is adequate, affordable amounts of safe, clean, sustainable electric power throughout this century and beyond for all the world. No more carbon in the atmosphere caused by electricity generation! No more used fuel disposal issues! No more concern about million-year integrity at geologic repositories. No more concern about dwindling energy resources for electricity generation! Multiple replicas of the demonstration reactor and phase-out of fossil-fueled electricity generators would accomplish this purpose.

ACKNOWLEDGEMENTS

Data on Brazil electricity generation were obtained from various Energy Information Administration sources. Extremely helpful discussions were held with Jose Mauro Esteves dos Santos, Antonio Carlos de O. Barroso, Jorge Spitalnik, and Everton Carvalho, key individuals in Brazil's nuclear energy organizations. However, conclusions about the potential of Brazil as a venue for demonstration of Super PRISM technology are the author's alone.

This work was performed under the auspices of the U.S. Dept. of Energy by the Univ. of Calif., Lawrence Livermore National Laboratory under contract No. W-7405-Eng-48.

REFERENCES

1. World List of Nuclear Power Plants, *Nuclear News*, p 35-58, March 2000.
2. International Energy Outlook 2000– With Projections to 2020, Energy Information AdministrationReport, DOER/EIA-0484(2000), March 2000.
3. Walter, Carl E., Directions for Advanced Use of Nuclear Power in Century XXI, Proceedings of American Nuclear Society Global '99 International Conference on Future Nuclear Systems, Jackson Hole, WY, August 30-September 2, 1999.
4. Generation IV: Looking to the Future of Nuclear Power, U.S. Department of Energy, Office of Nuclear Energy, Science, and Technology, January 2000.
5. Discussion on Goals for Generation IV Nuclear Power Systems, from a Workshop held May 1-3, 2000, July 31, 2000.
6. Lindholm, Ingemar, Depleted Uranium: Valuable Energy Resource or Waste for Disposal?, Uranium Institute 21st Annual Symposium (1996).
7. Annual Energy Outlook 2000– With Projections to 2020, Energy Information Administration Report, DOER/EIA-0383(2000), December 1999.
8. Radiation Standards– Scientific Basis Inconclusive, and EPA and NRC Disagreement Continues, Report to the Honorable Pete Domenici, U.S. Senate, U.S. General Accounting Office, GAO/RCED-00-152, June 2000.
9. Feb 17, 18 1993 speeches: http://www.whitehouse.gov/library/index.html.
10. Boardman, Charles E., et al, A Description of the S-PRISM Plant, Paper 8168, Proceedings of ICONE 8, 8th International Conference on Nuclear Engineering, Baltimore, MD, April 2-6, 2000.
11. Boardman, Charles E., Allen. E. Dubberley, and Marvin Hui, Optimizing the Size of the Super PRISM Reactor, Paper 8003, Proceedings of ICONE 8, 8th International Conference on Nuclear Engineering, Baltimore, MD, April 2-6, 2000.
12. Dubberley, A. E., C. E. Boardman, K. Yoshida, T. Wu, Super PRISM Oxide and Metal Fuel Designs, Paper 8002, Proceedings of ICONE 8, 8th International Conference on Nuclear Engineering, Baltimore, MD, April 2-6, 2000.
13. Fessler, Daniel, Concluding plenary talk at the American Nuclear Society's Global '99 Conference on Nuclear Technology– Bridging the Millennia, Jackson Hole, WY, September 2, 1999.

ENERGY TECHNOLOGIES AND CLIMATE CHANGE: A WORLD AND EUROPEAN OUTLOOK

Domenico Rossetti di Valdalbero
European Commission, Research DG (ENERGIE)[1]

INTRODUCTION

Beside the traditional policy and measures and other economic instruments (taxation, standardisation, voluntary agreements, internalisation of external costs, flexible mechanisms,...), the purpose of this paper is to analyse the role of new energy technologies to struggle Climate Change and to limit the costs of greenhouse gas emissions reductions, to present the European public energy expenses and to show the potential of a political impetus to increase the share of a specific technology.

Essentially based on the results of the POLES and PRIMES models[2] for the first and the second part and from EUROSTAT and SAFIRE model[3] for the third part, this paper presents:

I) the current and future energy situation and its correlated CO_2 emissions at world level (2030 time horizon) and European Union level (2010 time horizon) in the "reference case";

[1] The views presented in this paper do not necessarily reflect the European Commission official positions and the author alone assume the responsibility for the contents. The results presented in this paper come essentially from the POLES model (IEPE and IPTS), the PRIMES model (NTUA) and the SAFIRE model (ESD) essentially developed under the non nuclear energy programme of the European Commission, Research DG.

[2] POLES is a world energy partial equilibrium model which can simulate energy demand, supply and price formation for the main fuels and energy forms. PRIMES is a model that simulates a market equilibrium solution for energy supply and demand in the European Union Member States.

[3] SAFIRE is a technical economic model for new technology prospects and cost benefit evaluations.

Global Warming and Energy Policy, Edited by Kursunoglu *et al.*
Kluwer Academic/Plenum Publishers, New York 2001

II) five scenarios (nuclear, coal, gas, renewables and pessimistic) based on different technico-economic characteristics for energy technologies and their consequences on CO_2 emissions and marginal cost of emission reduction (World, 2030);

III) the European technological action (5th RTD Framework Programme and national public energy RTD expenses) and the specific political impetus for renewable energy (Europe, 2010).

1. PRESENT AND FUTURE OF ENERGY AND CO_2 EMISSIONS

1.1. World Energy reference case for 2030

Today, world Total Primary Energy Supply (TPES) is approximately 9500 millions tons of oil equivalent (Mtoe). In terms of fuel share, oil is at the first place with +/- 35% followed by solids (coal) with +/- 24%, gas with +/- 20%, combustible renewable and waste (biomass) with +/- 11%, nuclear with +/- 7% and hydro with +/- 2%. In terms of regional shares of TPES, OECD countries account for +/- 53%, the former USSR for 10%, China 12%, the rest of Asia 11%, Africa +/-5 % and Latin America +/- 5%.

Table 1: KEY INDICATORS AND WORLD PRIMARY ENERGY SUPPLY (POLES model)

POLES - **REFERENCE** WORLD		1990	2000	2010	2020	2030		y.a.g.r. 2000-30
Population	Million	5 249	6 150	7 027	7 893	8 713		1.2
Per capita GDP	90$/cap	5 217	5 714	7 142	8 862	10 732	+	2.1
GDP	G$90PPP	27 383	35 138	50 187	69 945	93 514	=	3.3
Energy intensity of GDP	toe/M$90	313	266	229	209	192	+	-1.1
Primary energy	Mtoe	8 338	9 359	11 517	14 639	17 944	=	2.2
Carb intensity of energy	tC/toe	0.70	0.69	0.71	0.73	0.75	+	0.3
CO_2 Emissions	MtC	5 863	6 443	8 188	10 692	13 411	=	2.5
Primary Energy Supply	Mtoe							
Solids		2 205	2 206	2 997	4 160	5 528		3.1
Oil		3 246	3 664	4 303	5 133	6 033		1.7
Gas		1 703	2 085	2 710	3 657	4 484		2.6
Others		1 183	1 404	1 507	1 689	1 900		1.0
of which								
Nuclear		433	602	623	687	759		0.8
Hydro+Geoth		184	224	279	341	408		2.0
Trad.Biomass		412	401	340	291	251		-1.6
Other Renewables		155	177	265	370	481		3.4

For 2030, global energy supply is likely to double and reach +/- 18000 Mtoe. The proportion of different energy sources will remain stable and fossil fuels will continue to be largely predominant (+/- 90% of TPES). The amount of coal will more than double mainly due to the Asian market development. Oil will remain the biggest source of energy at the world level increasing by two-thirds, essentially pushed by the transport demand. Gas will emerge as a leading element of the fuel mix, especially for the power generation sector.

1.2. A European Union reference case for 2010

EU energy demand is expected to continue to grow (1% pa) in the next ten years even if at rates significantly smaller than in history. The energy system remains dominated by gradually increasing imported fossil fuels. Solid fuels are expected to decline and gas is by far the fastest growing primary fuel (from less than 20% of primary energy consumption in 1990 to 26% in 2010). The share of oil is projected to be relatively stable and becoming almost exclusively a fuel for transportation.

Table 2: EUROPEAN UNION PRIMARY ENERGY SUPPLY AND DEMAND (PRIMES model)

PRIMES – **REFERENCE** EUROPEAN UNION	1990		2000		2010	
Primary Energy Mtoe	**Supply**	**Demand**	**Supply**	**Demand**	**Supply**	**Demand**
Solids	209	301	110	207	86	182
Oil	117	545	165	606	129	655
Gas	137	221	204	338	191	401
Others of which						
Nuclear	181	181	223	223	227	227
Renewables	61	61	79	79	89	88
Hydro	22		27		27	
Biomass and Waste	37		47		53	
Other Renewables	0		5		9	
Total	**705**	**1309**	**781**	**1453**	**722**	**1553**
CO_2 emissions MtC	855		872		960	

1.3. A world CO_2 emissions Outlook for 2030

According to the POLES model, world CO_2 emissions will more than doubling in the next 30 years. The picture between industrialised and developing countries will be reversed. If OECD countries represent today about half of total energy-related CO_2 emissions and developing countries less than one third, it will be approximately the opposite in 2030. Less-industrialised Asia will emit about 30% of the world CO_2 emissions, Latin America and the rest of developing countries 20%. USA, European Union and Japan together will account for about 30% of the world CO_2 emissions.

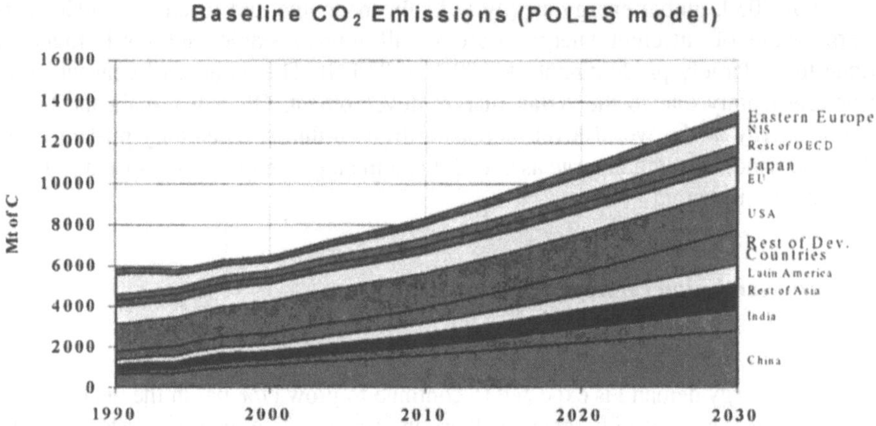

Graph 1: WORLD CO_2 EMISSIONS IN 2030 (POLES model)

This new balance in CO_2 emissions will result from differentiated demographic and economic dynamics but also from a movement of convergence in per capita CO_2 emissions among industrialised and developing countries. Nevertheless, with an average of +/- 4 tC per capita in 2030, the OECD inhabitants do not emit anymore seven times more but still four times more than their less-industrialised counterparts.

1.4. European Union CO_2 Emissions Outlook for 2010

After a short period of stabilisation at the beginning of the 90's, observed data (since 1994) and business as usual estimates show that CO_2 emissions are expected to increase by some 8% in 2010 from 1990 levels. Then, even if energy intensity improvements act in favour of moderating the rise of CO_2 emissions, the real effort for

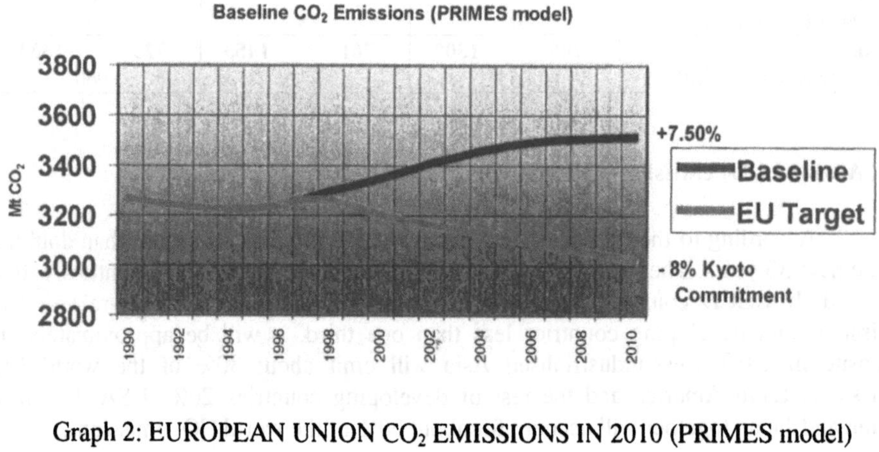

Graph 2: EUROPEAN UNION CO_2 EMISSIONS IN 2010 (PRIMES model)

Table 3: SECTORAL BREAKDOWN (1990 AND 2010) AND INCREASE OF CO_2 EMISSIONS IN THE E.U. (PRIMES model)

	1990	2010	INCREASE
ELECTRICITY HEAT	32%	30%	+ 2%
TRANSPORT	24%	30%	+ 39%
HOUSEHOLDS	21%	20%	+ 4%
INDUSTRY	19%	15%	- 15%
ENERGY BRANCH	4%	5%	+ 12%

the EU to achieve the Kyoto target (- 8%) amounts to a reduction of more than 15% (500 Mt of CO_2). The sectors with the fastest increase in emissions are those where energy demand is expected to grow fastest, namely the tertiary and transportation sectors. This latest, increasing by 39% in 2010 from 1990, will account for nearly two-thirds of the overall increase in emissions up to 2010.

2. TECHNOLOGICAL SCENARIOS AND COST OF EMISSION REDUCTION

2.1. Five technological scenarios (2030)

The five technological scenarios consisted in the identification of possible technological breakthroughs affecting clusters of technologies that are generically linked among themselves. Using expertise in technological perspectives, these breakthroughs are translated into more optimistic trajectories for the technico-economic characteristics of specific clusters of technologies. The POLES energy model is used to simulate in a consistent framework the impact of the scenarios by comparing to a reference case[4].

- The nuclear scenario assumes a breakthrough in nuclear technology in terms of cost and safety. Has an influence on standard large LWRs but especially on a new evolutionary nuclear design.
- The clean coal scenario involves major improvements in solid fuel burning technologies and especially affects the technico-economic characteristics of Super Critical Coal Plants, Integrated Gasification Combined Cycles and Advanced Thermodynamic Cycles.

[4] In affecting the simulations care was taken to reflect as completely as possible the competition between the various technologies, changed supply configurations and secondary effects regarding fuel prices.

- The gas technology scenario assumes enhanced availability of natural gas and introduces major technico-economic improvements for gas turbine combined cycles and combined heat and power plants. Some improvements assumed for all gas turbine related technologies. It additionally involves a breakthrough in fuel cell technologies in the latter part of the forecast horizon affecting proton exchange membrane fuel cells for fixed applications, solid oxide fuel cell with cogeneration and a hydrogen fuel cell car.
- The renewable scenario implying a major effort and breakthroughs in Renewable technologies, notably wind power, biomass gasification, solar thermal power plants, small hydro and photovoltaic cells.
- A generalised pessimistic scenario assuming rather unrealistically that the technico-economic characteristics of the technologies examined are frozen at their 1998 values for the duration of the forecast horizon (an exception was made for the standard LWR which in the reference case involved deterioration over time). The main purpose of this scenario was to examine the significance of technical progress already embodied in the reference case.

Table 2 illustrates a summary of the consequences of the technological scenarios on CO_2 emission and on the changes in the electricity production.

Table 4: IMPACT OF TECHNOLOGY SCENARIOS (World 2030)
Comparison to reference case in % (POLES model)

	Coal	Oil	Gas	Nuclear	Hydro	CO_2 Emmissions
Scenario:						
Nuclear	-9.9	-0.4	-2.3	116.7	-0.5	-5
Clean Coal	2.7	-0.8	-2	-9.9	-0.8	0.5
Gas & Fuel Cells	-18	-3.6	32.6	-16.5	-2.9	-2.2
Renewables	-9.5	-0.9	-3.2	-7.2	-0.8	-5.2
Pessimistic	0.9	1.9	3.7	16.9	-1.8	1.8

2.2. Cost of emission reduction with technology scenarios (2030)

For 2030, as a policy scenario an extension of the Kyoto type targets with full emission permit trading was assumed. The main assumptions of this scenario, which is radically different from the Business as Usual (BAU) case and is used as the reference for the present exercise are detailed briefly below.

- For Annex B countries other than those of the Former Soviet Union, the Kyoto targets were identically extended to the period 2010-2030.

- For countries of the Former Soviet Union which are in Annex B, a problem arises because according to the BAU scenario, their emissions decline by about 25% between 1990 and 2010 (instead of the zero growth Kyoto stipulation), thus giving rise to so called "hot air". For the purpose of the present exercise, their 2030 target was assumed to be stability at the 1990 value.

- As far as the rest of the World (non Annex B) countries are concerned, for which the Kyoto conference assigned no targets for 2010, we assumed that such targets would become operative beyond 2010 in such a way as to assure that World CO_2 emissions (including those targeted for Annex B countries) would stabilise at some level between 2020 and 2030.

Linking these assumptions with the technological scenarios, the role of technology in terms of marginal cost of reduction appears clearly in the Graph 1. In 2030, at the world level with a renewable or a nuclear scenarios the marginal cost of emission reduction passes from 175 \$/tC to respectively 140 and 120 \$/tC.

3. EUROPEAN RESARCH ACTION AND THE RENEWABLES PROGRESS

3.1. The 5th EU RTD Framework Programme

The 5th RTD FP[5] is a multiannual framework programme (1998-2002) for all Community activities, including demonstration activities, in the field of research and technological development. It covers four thematic programmes (Quality of life and management of living resources; User-friendly information society; Competitive and sustainable growth; Energy, environment and sustainable development) and three horizontal themes (Confirming the international role of Community research; Promotion of innovation and encouragement of participation of SMEs; Improving human research potential and the socio-economic knowledge base). Moreover, direct research activities are conducted by the Joint Research Centre.

From a total budget of M€ 14960, 7% are dedicated to energy in the "Energy, Environment and Sustainable Development programme" with the two key actions entitled "Cleaner energy systems, including renewables" (M€ 479) and "Economic and efficient energy" (M€ 547) and the RTD activities of a generic nature (M€ 16).

The 5th RTD FP includes Shared-cost actions (RTD, demonstration and combined projects, support for access to research, SME co-operative research projects and the SMEs exploratory awards), Training fellowships, Research training networks and Thematic networks, Concerted actions and Accompanying measures.

The EU Member States, the Central and Eastern European countries candidates for EU membership, the EFTA-EEA countries and Israel (and Switzerland when agreement will entry into force) can participate to the thematic programmes with Community funding.

The Community R&D expenses for energy have gradually decreased from the beginning of the 80's and have been stable in the 4th and 5th RTD FP.

[5] Decision n° 182/1999/EC of the European Parliament and of the Council (OJEC, n° L 26, 01/02/1999).

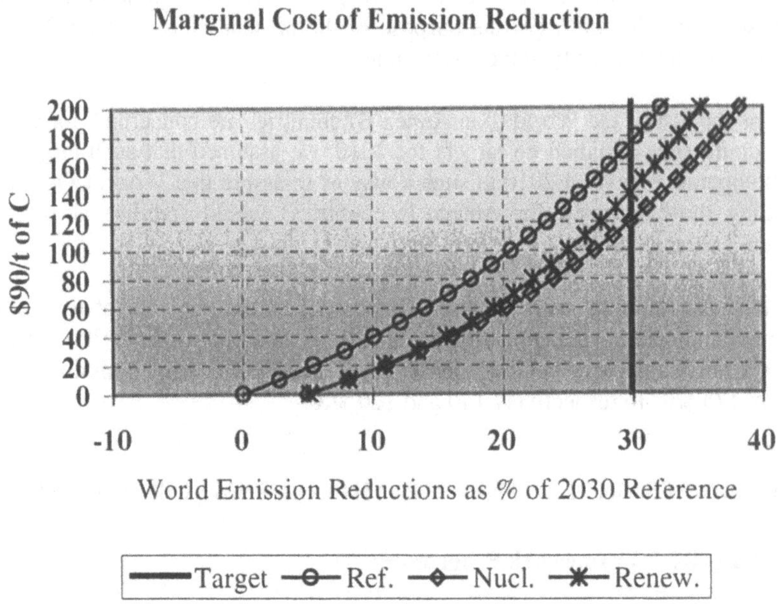

Graph 3: MARGINAL COST OF EMISSION REDUCTION (POLES model)

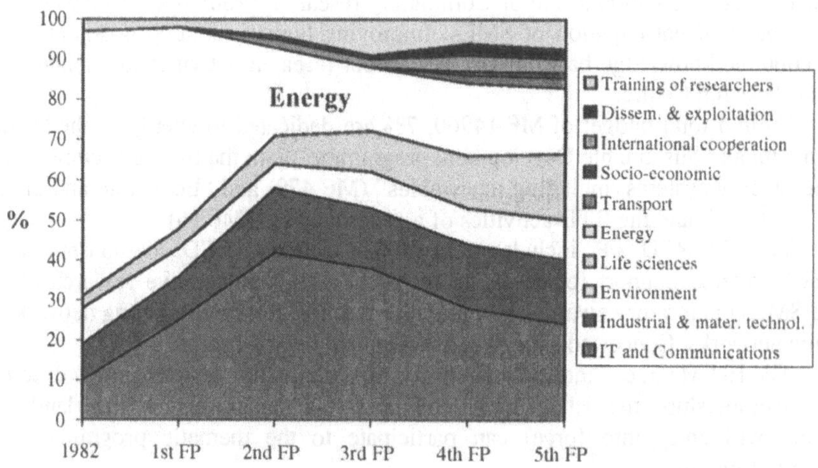

Graph 4: EU RTD CHANGING PRIORITY

3.2. The 15 E.U. National Renewable Energy RTD expenses

In these 10 last years, national energy RTD budgets are decreasing in most of the Member States. For renewables (except Austria, Belgium. Denmark, Finland, Greece, Netherlands), the reductions of government RTD budgets are particularly important in United Kingdom, Greece, Sweden, Italy and France.

According to the SENSER project[6], this is mainly due to:

- The liberalisation of energy markets and privatisation of energy industries;
- The decline in the perceived importance of energy security due to the relatively low energy prices and increased international energy trade;
- The progression towards maturity for certain technologies and an associated switch from RTD to demonstration and deployment measures.

The 15 EU Member States expenses for total solar RTD was 163 M$ (90 for PV) in 1988 and 110 (78 for PV) in 1998, for wind respectively 66 and 54 M$, for biomass 57 and 52, for geothermal 23 and 6, and for ocean 5 and 1.3. Only budget dedicated to hydro increased from 0.16 to 2 M$.

Table 5: E.U. MEMBER STATES RENEWABLES RTD EXPENSES (IEA statistics)

Government Energy Renewables R&D Expenses (M$95)			
	1988	**1995**	**1998**
Austria	5.8	10.7	10.5
Belgium	2.4	4.9	3.3
Denmark	4.9	19.5	18.7
Finland	0.00	7.1	12
France	9.7	6.2	4.1
Germany	91.5	96.2	81.4
Greece	32.7	3.5	5.9
Ireland	1.4	0.5	0.5
Italy	68.8	42	32.2
Netherlands	19.8	24	28.9
Portugal	2.8	0.6	0.5
Spain	16.6	16	13.1
Sweden	26	14.4	8.9
United Kingdom	32.1	14.4	5.5
Total EU15 MS	**315**	**260**	**226**
Total IEA Countries	624	683	568

[6] The SENSER project was co-financed by the European Commission, Research DG, non nuclear energy programme.

3.3. The European Policy Impetus for Renewables

Energy production from renewable energy sources (RES) in the EU has risen by about 25% between 1989 and 1998. The EU is at the forefront of the wind and PV industry in the rapidly growing world market of renewables. The European Commission[7] has fixed an indicative target of doubling the share of renewables passing from 6 to 12% of the gross inland energy consumption by 2010, i.e. achieving 22% of the electricity produced from RES.

Wind energy has been growing very fast in the 90's and accounting for 13.5 GW in the world and more than 6 GW in the EU (especially in Germany, Denmark and Spain) in 1998. It is expected to achieve 40 GW in Europe in 2010. Biomass is also increasing from 37 Mtoe in 1990 to 54 Mtoe in 1998 and it is scheduled to attain 135 Mtoe in 2010. PV remains very limited even if its installed capacities has been multiplied by 14 from 1990 to 1998 and it is expected to achieve 3 GW in 2010.

Table 6: RENEWABLE ENERGY INSTALLED AND EXPECTED CAPACITY IN THE E.U. (EUROSTAT and SAFIRE model)

Installed and expected capacities (European Union)				
MW electric	**1990**	**1995**	**1998**	**2010**
Hydro Without pumping	81,834	85,652	83,574	105,000
PV	7.6	32.5	110	3,000
Wind	483	2472	6132	40,000
Biomass	37 Mtoe	45 Mtoe	54 Mtoe	135 Mtoe
Solar panels	3.5 Million m^2	6.5 Million m^2	8.5 Million m^2	100 Million m^2

[7] Proposal for a directive of the European Parliament and of the Council *on the promotion of electricity from renewable energy sources in the internal electricity market* (OJEC, COM(2000)279 from the 10 of May 2000). Communication from the Commission, White Paper for a Community strategy and action plan, *Energy for the future: renewable sources of energy* (OJEC, COM(97)599 from the 27 of November 1997).

CONCLUSIONS

- In a reference case, the world energy demand and CO_2 emissions will approximately double in the next 30 years and developing countries will become major actors in the energy-environment field (cf. Table 1 and Graph 1). The EU situation will remain relatively stable in the next 10 years but with an increasing dependence from fossil fuels and, essentially due to more transport demand, CO_2 emissions will continue to grow (cf. Table 2, Graph 2 and Table 3).

- The Kyoto Protocol (December 1997) legally binding greenhouse gas emissions reduction targets (-8% for the EU, -7% for the USA and –6% for Japan) will require important efforts to be attained at the (2008-2012) "deadline". Different policy and market instruments can be used: the classic regulations or standards, the voluntary (environmental) agreements preferred by industry, the new flexible mechanisms, the highly discussed fiscal measures such as the carbon tax or, the topic of this paper, the RTD policy and the technological progress.

- Improvements in energy technologies ("technological breakthroughs") could reduce CO_2 emissions by +/- 5% in a nuclear or in a renewable scenario (cf. Table 4) in 2030 in comparison to a reference case. It thus appears that technological change by itself, i.e. with no accompanying measures, may have only a limited influence on the total greenhouse gas emissions (CO_2 emissions are expected to double up in 2030). On the other hand, in a "Carbon constrained world", assuming full play of "emissions trading", the marginal cost of abatement programs would be substantially reduced (by 20% to 30%) if breakthroughs in non-fossil technologies were achieved (cf. Graph 3).

- RTD policy is probably not the most effective but certainly one of the better-accepted policy to reduce CO_2 emissions. In fact, technological progress is generally well perceived by economic, political and social actors, cheaply implemented (a new administrative structure is not necessary and the transaction costs are very limited) and easily manageable on a European or world-scale (no national borders or legal constraints).

- Nevertheless, both at the European Community and Members States levels, the public RTD expenses for energy are decreasing (cf. Graph 4 and Table 5). For renewables, this reduction is particularly drastic in countries like the UK (six times less between 1988 and 1998), Greece (five times less), Sweden (tree times less), Italy and France (more than two times less). In the EU, with few fossil fuel resources, this reduction in RTD budget seems counterbalanced by the common European political will to double the share of renewables (cf. Table 6).

- More generally:

  * New technological breakthroughs appear as an essential instrument both at the world level to tackle the greenhouse gas emissions problem and, especially at the EU level, to alleviate the energy import dependency from risky geopolitical regions.
  * The on-going structural changes in the energy industry seem to strongly disadvantage the nuclear and renewables options but it appears that growing global environmental concerns (Kyoto) may herald a "new deal" for energy technologies in the next decades.

BIBLIOGRAPHY

Capros P. and al., (1999), "Climate Technology Strategies 1 & 2", Physica-Verlag, Heidelberg.

Capros P., (2000), "Energy Technology Dynamics and Advanced Energy Systems Modelling", International Journal of Global Energy Issues, Inderscience Enterprises, United Kingdom.

European Commission, (1999), "European Union Energy Outlook to 2020", European Communities, Belgium.

European Commission, (2000), "Proposal for a directive on the promotion of Electricity from Renewable Energy Sources in the Internal Electricity Market", COM(2000)279, Luxembourg.

Eurostat, (1999), "Integration – Indicators for energy", Official Publications of the European Communities, Luxembourg.

Lucas N., Zaleski, C., Rossetti di Valdalbero D., (1998), "Nuclear Scenarios and Related Policy Issues", in "Nuclear in a Changing World", EC - DG Research, Luxembourg.

Virdis M., Friedman K., Woodruff, M., (1998), "Fostering technological change: critical to the global energy future", World Energy Council, United Kingdom.

GLOBAL WARMING: A SCIENCE OVERVIEW

Michael C. MacCracken*

Fossil fuels (i.e., coal, oil, and natural gas) provide about 85% of the world's energy, sustaining the world's standard-of-living and providing the power for transportation. These fuels are inexpensive, transportable, safe, and relatively abundant. At the same time, their use contributes to problems such as air quality and acid rain that are being addressed through various control efforts and to the problem of global warming, which is now being considered by governments of the world. This paper describes six key aspects of the scientific findings on global warming.

1. HUMAN ACTIVITIES ARE CHANGING ATMOSPHERIC COMPOSITION AND INCREASING THE CONCENTRATIONS OF RADIATIVELY ACTIVE (GREENHOUSE) GASES AND PARTICLES

Observations from global measurement stations and reconstructions of the composition of the atmosphere in the past clearly indicate that human activities are increasing the atmospheric concentrations of carbon dioxide (CO_2), methane (CH_4), nitrous oxide (N_2O), and of various halocarbons (HCFCs and, until very recently, CFCs). These gases are collectively referred to as greenhouse gases because of their warming influence on the climate. The history of emissions versus concentrations, analyses of carbon isotopes, and other scientific results all confirm that these changes are occurring as a result of human activities rather than because of natural processes. The CO_2 concentration of almost 370 parts per million by volume (ppmv) is now about 30% above its preindustrial value of about 280 ppmv. The CH_4 concentration is up over 150%. While these gases occur naturally, records going back many thousands of years indicate that the present concentrations are well above natural levels. The concentrations of many halocarbons are entirely new to the atmosphere—many of these compounds are solely a result of human activities. The lifetimes of the excess contributions of these gases in the atmosphere range from decades

*Michael C. MacCracken, Executive Director, National Assessment Coordination Office, Office of the U. S. Global Change Research Program, 400 Virginia Avenue SW, Suite 750, Washington DC 20024 (on assignment from Lawrence Livermore National Laboratory); email: mmaccrac@usgcrp.gov; Web: http://www.nacc.usgcrp.gov.

Global Warming and Energy Policy, Edited by Kursunoglu *et al.*
Kluwer Academic/Plenum Publishers, New York 2001

(for CH_4) to centuries (for CO_2 and some halocarbons) to thousands of years (for some perfluorocarbons).

Human activities are also contributing to an increase in the atmospheric concentrations of small particles (called aerosols), primarily as a result of emission of sulfur dioxide (SO_2) from coal combustion. Once in the atmosphere, SO_2 is transformed into sulfate aerosols that create the whitish haze common over and downwind of many industrialized areas. This haze tends to exert a cooling influence on the climate by reflecting away solar radiation. Of critical importance is that the typical lifetime of aerosols in the atmosphere is less than 10 days (they are rained out as acid rain), so it is hard for global concentrations to build up in a way equivalent to the greenhouse gases with their longer atmospheric lifetimes.

Although natural processes can also affect the atmospheric concentrations of gases and aerosols, observations indicate that this has not been an important cause of changes over the past 10,000 years. Thus, it is well-established that human activities are the major cause of the dramatic changes in atmospheric composition since the start of the Industrial Revolution.

2. INCREASING THE CONCENTRATIONS OF GREENHOUSE GASES WILL WARM THE PLANET AND CHANGE THE CLIMATE

From laboratory experiments, from study of the atmospheres of Mars and Venus, from observations and study of energy fluxes in the current atmosphere, and from reconstructions of past climatic changes and their causes, it is very clear that the concentrations of key greenhouse gases play a very important role in determining the surface temperature of the Earth and other planets. Of the solar radiation reaching the top of the atmosphere, about 30% is reflected back to space by the atmosphere (primarily by clouds) and the surface; about 20% is absorbed in the atmosphere (primarily by water vapor, clouds, and aerosols), and about 50% is absorbed at the surface. As for all systems, the energy absorbed is then radiated away as heat (infrared radiation) based on the temperature of the object. Were the Earth's surface and atmosphere a simple radiator with the reflectivity of the present Earth, the average temperature would be very near 0°F (255 K), which would be much too cold to sustain life as we know it.

However, as heat is radiated from the surface, the greenhouse gases in the atmosphere absorb most of it. Less than 10% of the energy radiated from the surface gets through directly to space without being absorbed. A significant fraction of the absorbed energy is then radiated back to the surface, providing additional energy to warm the surface. This in turn causes more radiation to be radiated upward and again absorbed, providing more energy to be radiated back to the surface. This emission-absorption-reemission process is popularly called the *greenhouse* effect.

An additional warming influence results because the atmospheric temperature decreases with altitude up to the tropopause (about 8-10 miles up) before temperatures start to rise again in the stratosphere due to solar absorption by ozone. As more greenhouse gases are added, the absorption and back radiation to the surface comes from lower and warmer layers in the atmosphere, strengthening the greenhouse effect.

The greenhouse effect of the gases already mentioned is exceeded by the greenhouse effect of water, which is transported into the atmosphere through evaporation at the surface (and with a warmer surface temperature, more water vapor is lofted into the atmosphere). The water vapor condenses, which leads to formation of clouds; the condensation also releases heat into the atmosphere that is radiated both upwards and downwards, amplifying the greenhouse effect. Clouds both reflect solar radiation to space (exerting a cooling influence at the surface) and absorb and reradiate energy emitted upward by the surface (creating a warming influence at the surface).

Together, the natural greenhouse effect raises the average surface temperature of the Earth from about 0°F to almost 60°F (288 K). There is no scientific disagreement that if the addition of greenhouse gases to the atmosphere will tend to raise the average surface temperature. While aerosols exert a cooling influence, it would take an unrealistically large amount of aerosols to cause global cooling instead of warming as a net result of the use of fossil fuels.

3. INCREASES IN THE CONCENTRATIONS OF GREENHOUSE GASES SINCE THE START OF THE INDUSTRIAL REVOLUTION ARE ALREADY CHANGING THE CLIMATE, CAUSING GLOBAL WARMING

With the evidence indicating that the concentrations of greenhouse gases have risen significantly since the start of the Industrial Revolution and with the expectation that increasing the concentrations of greenhouse gases will cause warming, a key test of scientific understanding is to see if climatic changes are already occurring as a result of past emissions and if they are about of the magnitude that are expected based on theoretical and numerical analyses. Instrumental records of average temperature for large areas of the Earth go back to the mid-19th century. These records indicate a warming of over 1°F over this period. Extensive proxy records (i.e., records derived from tree rings, ice cores, coral growth, etc.) for the Northern Hemisphere going back about 1000 years also indicate very significant warming during the 20th century compared to the natural variations over earlier centuries. Thus, a sharp rise in the temperature seems to be occurring, and it is different in character than the earlier fluctuations that were likely caused by natural variations in solar radiation and the occasional eruption of volcanoes. That warming is occurring is also confirmed by rising temperatures measured in boreholes (i.e., dry wells), retreating mountain glaciers and sea ice, increasing concentrations of atmospheric water vapor, rising sea level due to melting of mountain glaciers and thermal expansion in response to recent warming (augmenting the natural rise due to the long-term melting of parts of Antarctica), and related changes in other variables.

The key question is whether these changes might be a natural fluctuation or whether human activity is playing a significant role. Among the reasons that the effect is being attributed largely to human activities is the coincidence in timing with the changes in greenhouse gas concentrations, the very large and unusual magnitude of the changes compared to past natural fluctuations, the warming of the lower atmosphere and cooling of the upper atmosphere (a sign of a change in greenhouse gas concentrations rather than in solar radiation), and the global pattern of warming. Some uncertainty is introduced because some of the warming occurred before the sharpest rise in greenhouse gas concentrations (probably a result of an increase in solar radiation and perhaps changes in land cover) and to the rise in tropospheric temperatures over the past two decades being a bit slower than the rise in surface temperatures (apparently a result of the confounding influences of ozone depletion, volcanic eruptions, and El Niño events).

Taking all of the scientific results into consideration, the Intergovernmental Panel on Climate Change (IPCC, 1996a) concluded in 1995 that "The balance of evidence suggests a discernible human influence on the global climate." This conclusion, in essence, is equivalent to the criterion for a civil rather than a criminal conviction. Since their 1995 report, the evidence has grown considerably more convincing, more clearly indicating that the magnitude and timing of the warming during the 20th century quite closely matched what would be expected from the combined influences of human and known natural influences.

4. FUTURE EMISSIONS OF GREENHOUSE GASES AND AEROSOLS WILL LEAD TO SIGNIFICANT FURTHER CLIMATE CHANGE, INCLUDING MUCH MORE WARMING AND SEA LEVEL RISE

With 6 billion people living on Earth and current average fossil fuel use, each person is responsible, on average, for emission of about 1 metric ton (tonne) of carbon per year. Per capita use varies widely across the world, reaching over 5 tonnes per year in the US and 3 tonnes per year in the OECD countries, but amounting to only about 0.5 tonne per person per year in developing countries. Projections for the year 2100 are that the global population may increase to perhaps 8 to 10 billion, and that, without emissions limits, average per capita emissions across the globe may double as fossil fuel use grows significantly in the highly populated, but currently underdeveloped, emerging economies. If this happens, total annual emissions would more than triple from about 6 billion tonnes of carbon per year to about 20 billion tonnes of carbon per year. New estimates by the IPCC suggest emissions in 2100 could range from about 6 to 60 billion tonnes carbon per year (IPCC, 2000).

Gradually increasing emissions that lead to emission of about 20 billion tonnes of carbon per year would raise the atmospheric CO_2 concentration to just over 700 ppmv. This would be almost double its present value, or over 250% above its preindustrial value. Projections based on the types of past changes that have occurred, on theoretical analyses, on understanding of planetary atmospheres, on extrapolation of recent trends, and, especially, from numerical climate models all suggest that this will lead to significant future warming. The 1995 IPCC assessment projected a global warming ranging from about 2 to over 6°F by 2100. If the world's nations control emissions of SO_2 (a step that seems necessary for health-related reasons), the increase by 2100 could be up to about 8°F.

Based on these projections, it is very likely that the world will warm significantly over the next several decades, even were there to be sharp reductions in CO_2 emissions. Associated with this warming would be shifts in precipitation zones, intensification of evaporation and precipitation cycles that are often associated with extremes of floods, droughts, and storms, and a significant acceleration in the rate of sea level rise. There are likely to be surprises as well, given the presence of potential thresholds and nonlinearities (as was the case for the Antarctic ozone hole). Among the possibilities are potential disruption of the Gulf Stream and the larger scale deep ocean circulation of which it is a part, the weakening of which apparently occurred as the world emerged from the last glacial about 11,000 years ago, causing a strong cooling centered over Europe (were this to happen during the 21st century, the cooling would likely only moderate somewhat the influence of global warming).

5. THE CONSEQUENCES OF CLIMATE CHANGE ARE LIKELY TO BE DIVERSE AND DISTRIBUTED, WITH BENEFITS FOR SOME, DAMAGES FOR OTHERS

With fossil fuels providing important benefits to society, contemplating changes in the ways in which most of the world's energy is generated would seem appropriate only if the types of consequences with which societies will need to cope and adapt are also quite significant. Several types of consequences for the US have been identified (NAST, 2000a; NAST 2000b) and the Intergovernmental Panel on Climate Change is summarizing impacts around the world (IPCC, 1996b, 1996c, 2001). The following sections summarize broad categories of impacts, particularly for the US which may be able to more readily adapt to these changes than those in developing nations.

5.1. Human Health

Sharp increases in summertime heat index may increase mortality rates unless offset by more extensive air-conditioning. The poleward spread of mosquitoes and other disease vectors may increase the incidence of infectious disease unless more aggressively managed by public health and building design measures. The increased intensity of extreme events may injure or kill more people (and disrupt communities) without more risk-adverse planning and construction.

5.2. Food Supplies

Increased CO_2 will aid growth of many crops and improve their water use efficiency. If this happens widely (i.e., if other constraints on agriculture do not arise), crop production should rise, increasing overall food availability, and reducing food costs for the public. For the farmer, the lower commodity prices that would be likely to result would stress farm income, and farmers in marginal areas, even though benefiting from some gain in productivity, are unlikely to remain competitive, causing economic problems in nearby rural communities unless other profitable crops are identified.

5.3. Water Supplies

Changes in the location and timing of storms will alter the timing and amount of precipitation and runoff, requiring changes in how water management systems are operated. This will be especially the case in the western US because there will likely be less snow and more rain in winter coupled with more and earlier melting of snow. These changes will likely require a lowering of reservoir levels in winter to ensure a greater flood safety margin, even though this risks reducing water availability in summer when demand will be rising. Increased summertime evaporation may also reduce groundwater recharge in the Great Plains, and cause lower levels in the Great Lakes and in rivers such as the Mississippi, stressing water transportation and recreation.

5.4. Fiber and Ecosystem Services from Forests and Grasslands

While winter precipitation may increase, temperatures will significantly increase. This is likely to reduce summertime soil moisture. Some, but not all, of these effects may be offset by the increased CO_2 concentration that will help many types of plants grow better (if other factors are not limiting). As seasonal temperatures and soil moisture change, ecosystems will be affected, causing changes in prevailing tree and grass types and then associated changes in wildlife. As regions accumulate carbon in vegetation and dry up, fire risk will increase in many regions. Some climate model projections suggest a much drier southeastern US, stressing the current forests. At the same time, the southwestern deserts may get wetter and sprout more vegetation (which may, in turn, increase fire risk in dry seasons). What is most important to understand is that the notion of ecosystem migration is a misconception—particular species will indeed grow in different locations. However, this will likely mean the tearing apart of existing ecosystems and the creation of new ones, albeit likely not with the complexity and resilience of current systems because there will not be sufficient time for adjustment and evolution to take place. If the climate changes as projected, stresses on ecosystems over the next 100 years may be as great as over the last 10,000 years.

5.5. Coastal Endangerment

Mid-range projections suggest that the relatively slow rate of rise of sea level this century (about 4 to 10 inches, reduced or amplified by regional changes) may increase by a factor of 3 during the 21st century. For regions currently subsiding (e.g. Louisiana, the Chesapeake Bay, etc.), there could be a significant acceleration in inundation and loss of coastal lands, especially of natural areas such as wetlands and other breeding grounds where protective measures such as diking cannot be afforded. The concern is greatest during coastal storms when storm surges (and therefore damage) will reach further inland and further up rivers and estuaries. For developed areas, strengthening of coastal protection is needed, not just to protect against sea level rise, but also to reduce current vulnerability to coastal storms and hurricanes.

5.6. Transportation

While the US transportation system is very reliable and quite robust, impacts from severe weather and floods currently cause disruptive economic impacts and inconvenience, even sometimes becoming quite important for particular regions. While information is only starting to emerge about how climate change might lead to changes in weather extremes, a range of possible types of impacts seem possible, including some that are location dependent and some that are event specific. Location-dependent consequences might, for example, include: lower summertime levels in the Great Lakes and the St. Lawrence and Mississippi-Missouri-Ohio River systems that could inhibit shipping; reduction in the thickness and duration of lake ice in the Great Lakes during winter that could lengthen the shipping season while also allowing higher winter waves to cause more damage to coastal infrastructure; sea level rise that could endanger barrier islands, and coastal infrastructure while shifting sediments and channels in ways that might affect coastal shipping and require more frequent remapping; and warming in the Arctic that could accelerate the reduction in sea ice, thereby opening the Northwest Passage for shipping while also causing more rapid melting of permafrost that could destabilize roads, pipelines, and other infrastructure. Event-specific consequences might, for example, include: more frequent occurrence of heavy and extreme rains (a trend already evident during the 20th century); reduced or shifted occurrence of winter snow cover that might reduce winter trucking and air traffic delays; altered frequency, location, or intensity of hurricanes accompanied by an increase in flooding rains; and warmer summertime temperatures that raise the heat index and may increase the need for air pollution controls. Early model projections suggest that the return period of severe flooding could also be significantly reduced (e.g., the baseline 100-year flood might occur, on average, every 30 years by 2100). Warmer temperatures will also reduce combustion efficiency, which would both increase costs and require longer runways or a lower load for aircraft. Starting to consider climate variability and change now in the design of transportation systems could be a very cost-effective means of enhancing both short- and long-term resilience.

5.7. Air Quality

Warmer temperatures tend to accelerate the formation of photochemical smog. The rising temperature and rising absolute (although perhaps not relative) humidity will raise the urban heat index significantly, contributing to factors that lead to breathing problems. Meeting air quality standards in the future is likely to require further reducing pollutant emissions (although, of course, a move away from the combustion engine may make this change much easier). Increasing amounts of ozone could also have greater impacts on stressed ecosystems, although the increased concentrations of CO_2 may help to alleviate some types of impacts. Summertime dryness in some regions could exacerbate the potential for fire, creating the potential for increased amounts of smoke, while in other regions dust may become more of a problem.

5.8. International Coupling

While it is natural to look most intently at consequences within a country, countries are intimately coupled to the world in many ways. For example, what happens outside the US will affect economic markets, overseas investments, the availability of imported food and other resources, and the global environment that all countries share. Health-related impacts overseas will be of importance to US citizens as travelers come to visit and as US citizens travel abroad for business and pleasure. Many resources, from water and hydropower-derived electricity to fisheries and migrating species, are shared across borders, or move and are transferred internationally. Finally, the US is largely a nation of immigrants, and when disaster strikes overseas, its citizens respond with resources and often by taking in refugees. Clearly, all countries are connected to what happens outside their borders.

5.9. Summary of Impacts

It is very difficult to accurately quantify the risk and importance of such a wide variety of impacts in a way that allows comparison with taking actions to change energy systems. At the international level, this is particularly difficult because issues of equity and cultural values more forcefully enter into consideration (e.g., what is the present economic value of the risk of the Marshall Islands being flooded over in 50 to 100 years?). Overall, there will likely be important consequences, some negative and some positive, that we are only starting to understand. Quite clearly, the present tendency to average across large domains covers over rather large consequences for smaller groups.

6. REDUCING THE RATE OF CHANGE OF ATMOSPHERIC COMPOSITION TO SLOW CLIMATE CHANGE WILL REQUIRE SIGNIFICANT AND LONG-LASTING CUTBACKS IN EMISSIONS

In recognition of the potential for significant change, the nations of the world in 1992 agreed to the Framework Convention on Climate Change (FCCC), which set as its objective the "stabilization of the greenhouse gas concentrations in the atmosphere at a level that would prevent dangerous anthropogenic interference with the climate system." At the same time, it called for doing this in a way that would "allow ecosystems to adapt naturally to climate change, ... ensure that food production is not threatened, and ... enable economic development to proceed in a sustainable manner." Defining the meanings of the terms and accomplishing the objective are both formidable challenges. For example, stabilizing the atmospheric concentration at double the preindustrial level (about 550 ppmv) would require stabilization of the present per capita CO_2 emission level at about 1 tonne of carbon per person per year rather than allowing it to double over the 21st century, as is projected to occur in the absence of controls (recall that the typical US citizen is now responsible for emission of about 5, European about 3, and the developing world individual about 0.5 tonnes of carbon per person per year). Then, for the 22nd century, global emissions would need to drop by at least a factor of 2 below current levels (i.e., to about half of the 6 billion tonnes of carbon now emitted each year).

Even though the Kyoto Protocol is controversial in the US, it would be only a rather modest step in starting to limit emissions. If fully implemented through the 21st century, the increase in the CO_2 concentration from present levels would be reduced about 15-20%, rather than the 50% cutback needed to achieve stabilization at 550 ppmv. Limiting the projected increase in emissions will require significant introduction of non-fossil energy technologies, improvement in energy generation and end-use efficiencies, and switching to natural gas from coal (or even worse from a CO_2 emissions standpoint, from oil-shale-derived energy). What is clear from present energy analyses is that there is no "silver bullet" that could easily accomplish a major emissions reduction. Achieving the FCCC objective over the 21st century is therefore likely to require an aggressive (but not unprecedented) rate of improvement in energy efficiency, broad-based use of non-fossil technologies (often selecting energy sources based on local resources and climatic conditions), and accelerated technology development and implementation.

7. CONCLUSION

A major reason for controversy about dealing with this issue results from differing perspectives about how to weigh the need for scientific certainty, about ensuring a reliable source of energy to sustain and improve the national and global standard-of-living, about capabilities for improving efficiency and developing new technologies, about the potential risk to "Spaceship Earth" that is being imposed by this inadvertent and virtually irreversible geophysical experiment, about the economic costs and benefits of taking early actions to reduce emissions (including what factors to consider in the analysis and how to weigh the importance of long-term potential impacts versus better defined near-term costs), and about the weight to give matters of equity involving relative impacts for rich versus poor within a nation, the developed versus developing nations, and current generations versus future generations.

I believe that coming to a consensus on these issues will require that we all become better informed and that collective action will require that the political system focus on finding approaches that tend to balance and reconcile these (and additional) diverse, yet simultaneously legitimate, concerns..

Acknowledgements: The views expressed are those of the author and do not necessarily represent those of his employer or the US Global Change Research Program. This paper was prepared under the auspices of the Department of Energy's Environmental Sciences Division by the Lawrence Livermore National Laboratory under contract W-7405-ENG-48.

REFERENCES

IPCC (Intergovernmental Panel on Climate Change), 1996a, *Climate Change 1995: The Science of Climate Change*, edited by J. T. Houghton, L. G. Meira Filho, B. A. Callander, N. Harris, A. Kattenberg, and K. Maskell, Cambridge University Press, Cambridge, United Kingdom, 572 pp.

IPCC (Intergovernmental Panel on Climate Change), 1996b, *Climate Change 1995: Impacts, Adaptations, and Mitigation of Climate Change: Scientific-Technical Analyses*, edited by R. T. Watson, M. C. Zinyowera, and R. H. Moss, Cambridge University Press, Cambridge, United Kingdom, 879 pp.

IPCC (Intergovernmental Panel on Climate Change), 1996c, *Climate Change 1995: Economic and Social Dimensions of Climate Change*, edited by E. J. Bruce, Hoesung Lee, and E. Haites, Cambridge University Press, Cambridge, United Kingdom, 464 pp.

IPCC (Intergovernmental Panel on Climate Change), 2000, *Special Report on Emissions Scenarios*, N. Nakicenovic (lead author), Cambridge University Press, Cambridge, United Kingdom, 599 pp.

IPCC (Intergovernmental Panel on Climate Change), 2001, *Climate Change 2001: Impacts, Adaptation, and Vulnerability*, edited by J. J. McCarthy, O. F. Canziani, N. A. Leary, D. J. Dokken, K. S. White, Cambridge University Press, Cambridge, United Kingdom, in press.

NAST (National Assessment Synthesis Team), 2000a, *Climate Change Impacts on the United States: The Potential Consequences of Climate Variability and Change, Overview*, U. S. Global Change Research Program, Cambridge University Press, Cambridge, United Kingdom, 154 pp. (Also see http://www.nacc.usgcrp.gov).

NAST (National Assessment Synthesis Team), 2000b, *Climate Change Impacts on the United States: The Potential Consequences of Climate Variability and Change, Foundation*, U. S. Global Change Research Program, Cambridge University Press, Cambridge, United Kingdom, in press (Also see http://www.nacc.usgcrp.gov).

"IS NUCLEAR ENERGY GOING TO MISS ITS ENVIRONMENT MISSION ?"

Juan Eibenschutz

Nuclear energy was developed in the twentieth century, it is the only "new energy", and in a relatively short time period it has reached the same level of participation in the world's energy balance of hydro energy, probably the first energy form discovered by man.

Fossil energy is by far the main supplier of energy, resources are plentiful, and although fossil energy resources are finite, they will provide the needs of the world for several decades into the future.

Energy is one of the main constituents of the present economy. Modern society takes it for granted, and the less developed countries are depending on energy to attain their development goals. Energy shortages can stop the functioning of towns, cities or countries, and this is why every country does whatever it deems necessary to assure their energy supplies.

Two factors are probably the most relevant in the present energy scenarios: market economy and the environment.

Because the main energy supplies come from oil, coal and gas, and since the use of these materials involves combustion, there is a pollution problem arising from the energy industries, as well as from the utilization of fuels.

On another dimension, the governments of the leading countries, as well as most of the governments of the developing world, have decided to let energy supplies in the hands of the market economy.

OECD countries have strategic energy policies, and while letting the market handle the tactics of supplies, they do whatever is necessary to guarantee the availability of supplies. The developing countries try to do the same, within the limitations derived from their economic strength, and the amount of their energy resources.

But, there is a new societal movement at the end of the twentieth century: subjectivism (or politics for the sake of politics). Science, pragmatism and objectivity are yielding to feelings, religious radicalism, fantasy and virtualism, forcing governments to react and commit unfeasible policies and unattainable objectives.

Luz y Fuerza del Centro, Mexico

Global Warming and Energy Policy, Edited by Kursunoglu *et al.*
Kluwer Academic/Plenum Publishers, New York 2001

This phenomenon can be observed in many fields, for example in the religion driven confrontations within countries and between some countries or, in the education goals set forth in some developing countries.

However, the energy field seems to be particularly affected by the phenomenon, some examples are the deregulation policies established for the electricity sector in several countries that had to be changed into re-regulation, because free competition was leading to unreasonable prices. Or the exaggerated reliance on combined cycle power plants, to satisfy growing electricity requirements with the new technology, that is not yet mature enough. Or in another example, the goals accepted in several countries for the participation of renewable energy sources that may solve some special energy needs, but cannot substitute for conventional energy sources.

One particularly disturbing example is the future level of greenhouse gases that has been accepted by many countries, in spite of the fact that new emission free sources will not be deployed in time to meet the goals.

And worse yet, the subjectivism that has been developing in many societies has affected nuclear power very seriously. We know that there are technical solutions to deal with all the safety and radioactive residues issues involved in the nuclear industry, but we also know that there are serious concerns in many societal groups, and that the public has acquired consciousness and will not accept impositions.

Is the anti-nuclear position similar to the rejection of medical assistance by some religions? Maybe, because although nuclear energy is not the only solution, its abandonment will unavoidably lead to more pollution, and possibly to severe changes in the conditions of survival on earth.

The issue of partiality against nuclear has received substantial lip service, but the problem is definitely one of subjectivity. Many people are against nuclear because they don't accept or don't understand the implications of their antagonism.

The present levels of well being in the world, regardless of the enormous differences between the haves and the have nots, have been achieved because society accepts risks, maybe because most people do not care to analyze alternatives, or because many risks are implicitly accepted.

Take for example the risks involved in taking medicines. The use of any medicine implies a certain risk, but the benefits largely override the risks, and therefore most people are willing to accept the risks.

Nuclear energy was first presented to society by the bomb, then the risks of radiation were publicized, therefore most people have a predisposition against nuclear. Regardless of the fact that the benefits of nuclear power are considerably higher than the risks, the perception is that the opposite holds true. What would have happened, as Dean Nabor Carrillo pointed out, if the first application of electricity would have been the electric chair?

Most likely, people would have not accepted electric wires to be installed in their homes.

To objectively show the position of nuclear power, a publication (Electric Power from Competitive Sources) by Lindsay Juniper, Managing Director of Ultra-Systems Technology Pty. Ltd., Brisbane, Qld. Australia, is useful. Juniper's work provides a simple computer program to compare the costs of electricity produced from different

energy sources, and provides a set of values obtained by him that allow a discussion on the subject.
For example, the following table from this work shows the environmental costs, associated to the different technologies:

Table 1. Summary of Environment Costs

Cost Item	Coal-Fired Plant	Gas-Fired CC	Nuclear
Air pollution control	6 – 18 %	0 – 6 %	
Cooling	0 – 2 %	0 – 3 %	
Environment charges	0 – 9 %	0 – 5 %	
SO_2 and NO_x control	15 – 20 %		
Particulate control	3 – 4 %		
Fuel disposal			1 – 4 %
Safety systems			15 – 45 %
Total	12 – 42 %	0 – 9 %	15 – 50 %

From this table, it is clear that environmental costs are comparable between nuclear and coal, except that these costs are always internalized for nuclear and not for the other sources.
The following table, also taken from the reference, shows the base costs from the different power sources.

Table 2.

Overall Summary Of Power Generation Costs									
		Fossil Fuels			Renewables				Nucle-ar
		Coal	Oil	Gas	Solar PV	Sol. Therm	Wind	Bio-mass	
Station Size	MWe	1,320	1,000	400	1.0	160	5	50	2,000
Number of Units		2	2	2	1	1	5	1	2
Annual Capacity Factor	%	85.0	90.0	90.0	15.0	20.0	40.0	75.0	80.0
Station Annual Output	GWh/a	9,147	7,495	2,872	1.2	258	16	302	13,324
Annual Capital Charges	US$/MWh	14.92	10.93	10.84	556.67	197.13	31.75	18.49	25.52
Annual O&M	US$/MWh	5.35	5.65	3.04	11.37	40.62	4.61	10.51	6.79
Annual Fuel Cost	US$/MWh	15.48	25.58	27.39				0.65	8.36
Bulk Electricity Supply Cost	US$/MWh	35.75	42.16	41.27	568.03	237.75	36.37	39.65	40.67

It is interesting to note that except for wind and biomass, that are not alternatives to supply the electricity needs on the scale required, the range is reasonably close. The following tables provide data on some of the factors that are relevant in the analyses of the possible effects of abandoning nuclear power.

Table 3. NO_x Emissions from Combustión Plants

Fuel Type	Uncontrolled Emissions	Controlled Emissions
Coal	400 – 1,200 ppm	150 – 400 ppm with Low-NO_x burners. Less than 100 ppm with SCR
Oil	100 – 200 ppm	Down to 50 ppm
Gas	50 – 100 ppm	Down to 10 ppm
Renewables: Solar & wind	Zero	
Renewables: Biomass	200 – 400 ppm	Generally no reduction mechanisms installed
Nuclear	Zero	

Table 4. CO_2 Emissions from Combustion Plants

Fuel Type	Specific CO2 Emissions (g/MJ)	Emissions from Plants (kg/MWh)
Coal	85 – 95 for bituminous coal	850 – 950 for subcritical units Down to 800 for supercritical
Oil	70 – 75	700 – 750
Gas	50 – 55	450 – 500
Renewables: Solar & wind	Zero	
Renewables: Biomass	90 – 100	>1,000 Note that combustion of biomass is considered CO2 neutral
Nuclear	Zero	

The two previous tables provide quantitative support to the argument of the very severe risks that abandoning nuclear power implies for the future of mankind.

Although climate change due to the greenhouse effect of combustion gases released into the atmosphere is not fully accepted by the scientific community, it is clear that the future energy requirements can only be satisfied with energy provided by sources capable of producing the enormous amounts required, particularly by the emerging countries, whose development is unthinkable without energy.

The two tables that follow show that the capacities available from the non-polluting sources, other than nuclear, make them unfeasible as alternatives.

Table 5. Workforce Requirements

Energy Source	Unit Output	Construction Workforce		O & M
		Peak	Total	Workforce
	MW	Number	Man years	Number
Fossil fuel energy				
Coal fired	500	400	1000	250
	100	200	450	150
Oil fired	500	300	600	200
	100	150	250	100
Natural gas fired	500	250	500	150
	100	125	200	75
Renewable energy				
Solar photovoltaic	1	30	10	5
	0.2	20	5	3
Solar thermal	1	40	15	8
	0.2	25	10	4
Wind	5	20	10	10
	0.5	15	5	5
Biomass	40	50	60	30
	5	30	25	15
Nuclear energy				
PWR/BWR	1000	1000	3000	300
PWR/BWR	500	700	2000	200

Table 6. Site and Infrastructure Requirements

Energy Source	Unit Output	ACF	Energy Output	Site Area	Comments
	MW	%	GWh/a	ha	
Fossil fuel energy					
Coal fired	500	85	3723	20	Includes coal stockpile, ash dam and switchyard
	100	80	701	10	
Oil fired	500	88	3854	12	Includes oil tankage and switchyard
	100	83	727	7	
Natural gas fired	500	89	3898	10	
	100	84	736	6	
Renewable energy					
Solar PV	1	20	1.8	1	Assume 12% conversion efficiency
	0.2	17	0.3	0.3	Assume 12% conversion efficiency
Solar thermal	1	22	1.9	1.5	Assume 15% conversion efficiency
	0.2	18	0.3	0.4	Assume 15% conversion efficiency
Wind	5	25	11	3	Depends on unit size and wind speed
	0.5	20	0.9		
Biomass	40	60	210	2	Includes biomass storage
	5	40	18	1	Includes biomass storage
Nuclear energy					
PWR/BWR	1000	95	8322	15	Includes buffer zone
PWR/BWR	500	93	4073	10	Includes buffer zone

Furthermore, the spatial requirements of renewables are a clear indication of the very low energy density of these technologies, that result in impossibly high costs.

From the reference the last table, that follows, is an interesting comparison between the different power technologies, its analysis is relevant for our subject, particularly because the author is obviously not biased in favor of nuclear.

The main factor affecting nuclear is political or societal; it has to do with public opinion and public pressure.

The following table summarizes the general outcomes of this study and provides a quick reference on the costs of electric power from competitive sources.

Table 7. Summary of Study Outcomes

	Coal	Oil	Gas	Renewables				Nuclear
				Solar PV	**Solar Thermal**	**Wind**	**Biomass**	
Plant Details								
Unit Size	200 – 1,300 MW		200 – 750 MW	0.01 – 1 MW	Up to 200 MW	Up to 1.5 MW	Up to 100 MW	600 – 1,300 MW
Capacity factor	75 – 95%	75 – 95%	75 – 95%	5 – 20%	5 – 20%	Highly site specific, but typically around 20 – 50%	50 – 80 depending on availability of fuel	60 – 90%
Fuel Source								
Abundance	Very abundant	Abundant bul controlled	Abundant in many locations	Very abundant but limited capacity factor	Very abundant but limited capacity factor	Abundant in many locations but limited capacity factor	Limited resources in most locations	Abundant
Security of supply	High	Risky due to political whims of unstable governments	Moderate Available from a number of limited sources	Subject to weather patterns	Subject to weather patterns	Subject to weather patterns	Subject to weather patterns to grow primary mass	Risky due to public pressure on environment issues
Cost	Cheap	Moderate	Expensive	Zero	Zero	Zero	Very cheap	Very cheap
Power Plant Costs								
Capital	Cheap – Moderate	Cheaper	Cheap	Very expensive	Expensive	Cheap	Moderate	Moderate – expensive

Table 7. (Continued)

	Coal	Oil	Gas	Renewables				Nuclear
				Solar PV	Solar Thermal	Wind	Biomass	
Operating & maintenance	Cheap	Cheap	Cheap	Moderate	Expensive	Cheap	Moderate	Moderate
Cost of electricity supply[2]	100 % Cheap	116 % Moderate	107 % Cheap	1,380 % Very expensive	725 % Very expensive	105 % Cheap	102 % Cheap	115 % Moderate
Environmental Issues								
Polluting potential	High	Moderate	Low	Very low	Very low	Very low	Low	Potentially very high
Difficulty of compliance	Hard but technology has developed rapidly	Moderate only due to high sulphur fuel oil	Easy	Easy	Easy	Easy	Moderate	Difficult
Cost of compliance	High	Moderate	Low	Very low	Very low	Very low	Low	High
Social/Political Issues								
Potential to fulfill community energy needs	Very high	Possible but not currently in favor	High based on fuel availability	Very low	Very low	Low	Low	High but subject to public opinion
Basis for reform of energy market	High	High but risk of fuel supply	High	Low	Low	Low	Low	High but subject to public opinion

The facts, highlighted by the previous discussion demonstrate that nuclear power competes on economic grounds with other sources, and that on environmental terms is the most benign source of power.

If subjectivity prevails nuclear power will be abandoned and the future of the global society will be seriously handicapped.

CALIFORNIA'S ELECTRICITY PROBLEM – AND THE POTENTIAL WORLD ENERGY DISASTERS

Bertram Wolfe

Is it surprising that California is now suffering from a lack of needed economic electricity? Last winter this happened in other parts of the country, and it is projected that this winter, with cold weather, there will again be energy shortages in various parts of the nation. Why is our welfare in jeopardy?

The answer is that this nation has not looked responsibly to the future. Does anyone remember that in the late sixties and early seventies we were doubling our energy use and supply every 10 years? Can one recall that, during this period, energy planners and environmental organizations like the Sierra Club were supportive of nuclear energy? With the needed energy increases, one had to choose between the various fossil fuels and nuclear power; and the Sierra Club felt that nuclear energy was least damaging to the environment.

In 1973 the situation changed. The Arab oil boycott, and the resulting higher energy costs, slowed down the growth of electricity in this country from a doubling every ten years to a doubling of some 35 years. As a result, the new plants ordered before 1973, and subsequently built, have led to a surplus of electrical supply in this country. Since 1973, then, the "environmental" organizations have been able to oppose all new energy plants. These include oil, gas, coal, and nuclear plants, as well as dams, and even geothermal plants. They argue for solar and wind power, which on a large scale are impractical, and indeed are environmentally detrimental. But with a surplus of energy supply, it doesn't matter.

The situation is now changed, and is becoming more desperate because we have let our electrical surplus vanish. We now need new energy capacity to meet our present needs, and to meet our continued electrical growth which still amounts to a doubling in 35 years. Further, approaching global problems may be devastating. These are caused largely by the increasing population of the third world, which is looking to raise its standard of living. The world population in the next 50 years is projected to increase from 6 to some 10 billion people and if the average world per-person energy use reaches only one third of that in the US today, then world use will triple.

Global Warming and Energy Policy, Edited by Kursunoglu *et al.*
Kluwer Academic/Plenum Publishers, New York 2001

This can result in major catastrophes. One is potential international hostilities over scarce energy supplies. (Recall that concern over needed energy was one of the reasons for Japan's entry into WWII.) Another is the potential global warming disasters caused by the use, and increased use, of fossil fuels.

The one available solution to these potential disasters is a major increase in the utilization of nuclear energy. Almost every technology has potential dangers. Consider the thousands of deaths from the 1984 chemical plant explosion at Bhopal, India; the fifty thousand automobile deaths each year in the US; the thousands of yearly US deaths from fossil fuel particle emissions; and even the danger from McDonald's hot coffee. Nuclear energy has its potential problems; but the nuclear plants in the Western world, including Three Mile Island, have not harmed a single member of the public. Chernobyl would not have been allowed here, and the Russians are now adopting Western safety standards. Similarly, nuclear wastes have not harmed anyone; and the problems with implementing waste repositories like Ward Valley and Yucca Mountain have been political, not technical, fostered by anti-nuclear organizations who frighten the public.

Nuclear energy can provide an essentially unlimited supply of energy. Anti-nuclear activists frighten the public about nuclear wastes thousands of years out. But the real potential catastrophe is lack of energy in the coming decades when oil and gas supplies are used up; and in the following century when economic coal supplies are depleted. The near term use of nuclear energy allows us to lengthen the availability of specially needed fossil fuels; and although long term nuclear wastes can be safely accommodated, advanced nuclear plant designs will allow us to modify the nuclear wastes so that they lose their radioactivity in just a few hundred years.

There is little question that, in the future, need will lead to a major world expansion of nuclear energy. The real question is whether unnecessary national and world suffering will occur because of delays in the sensible revival and world expansion of nuclear energy.

Today we are having very disturbing, but relatively mild, energy problems in the US due to our lack of preparation for the end of our energy surplus. Again, the question we must answer is whether we will wait for the potential national and world energy disasters to occur, or whether we will now expand nuclear energy to mitigate, or eliminate, them.

NUCLEAR ENERGY, NON-PROLIFERATION, AND OTHER CONSIDERATIONS

NON-PROLIFERATION ISSUES FOR GENERATION IV POWER SYSTEMS
Advanced waste management

Joseph Magill and Roland Schenkel*

1. INTRODUCTION

In the event of a revival of interest in nuclear energy, so-called Generation IV[1] power systems will come into operation between 2030-2050. These systems should be highly economical, have enhanced safety features, give rise to a minimum of waste, and be proliferation resistant.

Prominent among the performance goals of such reactors is the fact that they should be competitive with other electricity generating sources. Specifically, the cost in US dollars should not exceed 3 cents/kWh on the basis of year 2000 prices. In addition, plant capital investment costs should not exceed $1000/kWe on the same basis.

Another performance goal is that these Generation IV systems should have high proliferation resistance. The fear is sometimes expressed that, in a nuclear revival situation, where many different power systems may be available, one can no longer control the flow of fissile material and the uses to which such systems may be put (e.g. clandestine fissile material production). Taken to the extreme, can proliferation issues become "showstoppers"? In the following paper, we consider Generation IV systems from the proliferation viewpoint and describe activities at the Institute for Transuranium Elements in this area. The main subjects to be covered in this article are:

- Safeguards issues with Generation IV reactors:
- New Safeguards Issues within Advanced Waste Management
- Case Study: Accelerator Driven Systems (ADS)
- Exotic Nuclear Systems
- The Cost of Safeguards

* European Commission, JRC, Institute for Transuranium Elements, Postfach 2340, 76125 Karlsruhe, Germany

The potential proliferation risks that might arise from the implementation of partitioning and transmutation (P&T) scenarios need to be analysed to identify proliferation pathways, to select the most proliferation resistant option, and to implement the appropriate safeguards control mechanisms. Specifically, we consider potential proliferation issues associated with a) advanced aqueous and pyroprocessing of present nuclear fuel to process spent fuel, b) the separation and recovery of minor actinides Np, Am, Cm, and c) the transmutation of these actinides in accelerator driven systems.

In a Case Study on accelerator driven systems, we describe a particular example of how some potential proliferation problems were identified in ADS systems and how this has led to new regulations regarding the export of associated technology.

Traditional nuclear systems are based on the use of uranium, thorium and plutonium. Future sytems may be based on other nuclear materials such as neptunium and americium. These materials are not covered within the existing safeguards agreements. We describe some exotic applications based on the use of americium which may have a role to play in future nuclear systems.

Finally, it is also shown that the costs associated with safeguarding present day civilian reactors are very small indeed compared to the overall capital investment and operational costs.

2. SAFEGUARDS ISSUES WITH GENERATION IV REACTORS

Despite the lack of confidence by the public, nuclear energy is faced with developments which may overcome the world-wide stagnation in the medium term. These developments are not only due to the prolongation of plant lives in the US, but also due to initiatives on innovative designs of reactors/fuel cycles with the following objectives:

- Improved economic competitivity
- Increased nuclear safety
- Minimised wastes (in particular with long-lived radiotoxicity)
- Optimised fuel efficiency, use of all fissile/fertile resources
- Enhanced proliferation resistance and/or safeguardability

In addition to classical base load electricity production, there is an increasing need for high temperature process heat applications, hydrogen production and low temperature seawater desalinisation plants for water production in developing countries. Nuclear energy is one of the promising options for economic, reliable and CO_2 free production of such energy. The evolution of nuclear power as foreseen by the Generation IV initiative is shown in table 1.

Among the innovative reactors, the following are under consideration: advanced light water reactors as large or medium size plants; high temperature gas cooled reactors; liquid metal or gas-cooled fast reactors including modular sized (300 MWth) passively safe; molten salt based reactors; and accelerator driven systems (ADS).

The main inspection goals are to detect the diversion of a) one irradiated fuel assembly (or equivalent number of pins) within 3 months, b) one fresh fuel assembly within 12 months c) one MOX fuel assembly within 1 month with a detection probability greater than 90%.

Table 1. Evolution of Nuclear Power Reactors

Generation I:	Generation II:	Generation III:	Generation IV:
Early prototypes	Commercial Power Reactors	Advanced LWRs	Under-development NEWS reactors
Magnox	L W R	E P R	Non-proliferating
Calder Hall	Candu	ABWR	Economical
Heavy Water	VVER/RBMK	AP600	Waste minimisation
			Safe
1950	1970	2000	2020

2.1. Advanced Light Water Reactors

The new features of these reactors will have no impact whatsoever on the safeguardibility of the plants. Current practices of the IAEA, including the use of MOX fuel are state of the art and can be carried out to the full satisfaction of the IAEA.

2.2. High Temperature Gas Cooled Reactors

Also in the area of High Temperature Gas Cooled Reactors, operators and the safeguards inspectorates have considerable experience and know-how, as such reactors have been and are being safeguarded in different countries. Depending on design changes in fuel load and discharge mechanisms or fuel types, some additional and plant specific on-line verification systems may need to be developed. This could also apply to plutonium fuel usage in HTGRs. However the basic techniques like neutron-coincidence and gamma-ray measurement techniques are readily available.

2.3. Liquid Metal or Gas-Cooled Reactors

Several prototype reactors have been or are in operation in non-nuclear weapon states (Germany, Japan) and nuclear weapon States (United Kingdom, France, Russia). Operators and inspectorates (IAEA/Euratom safeguards) have established safeguards concepts and implemented successfully safeguards including the Superphenix reactor. Therefore, also for this type of reactor considerable experience and adequate equipment is available to achieve a very high degree of assurance by application of international safeguards.

Less experience is available with relation to gas-cooled reactors and new type of fuels (for example metal or nitride fuel). In such a case, however, available techniques could be adapted relatively easily to new matrices and geometries.

A new element would be the use of "cartridges" for reactors with refuelling periods of up to 15 years. Provided adequate verification can be performed on the fresh cartridge, a highly tamper resistant sealing mechanism could be applied which permits even remote control of integrity by an inspection agency.

2.4. Molten Salt based Reactors

These reactor concepts have their origins in Integral Fast Reactor developed at Argonne. The molten salt reactor is possibly the only reactor type under consideration, which requires a detailed diversion analysis and development of a safeguards concept. The same holds true for pyrochemical reprocessing. It should be noted, that both concepts are considered very proliferation resistant due to the fact, that fissionable isotopes are never handled in pure and separated form. At the Institute for Transuranium Elements (ITU), for example, we will develop suitable techniques to perform flow and inventory measurements of fissile materials in molten salt solutions (for operator material accountancy). As a spin-off, concepts for inspection verification will be developed.

2.5. Accelerator Driven Systems (ADS)

Over the past 10 years, accelerator driven systems have been proposed both for energy production and nuclear waste transmutation. First demonstrations plants may be in operation around 2010 but full reactors will not come into operation before 2030. Initial studies indicate that some new safeguards problems can be expected with such systems. A more detailed description of these are given later in this paper.

3. NEW SAFEGUARDS ISSUES IN ADVANCED WASTE MANAGEMENT SYSTEMS: PARTITIONING AND TRANSMUTATION

In this sixth decade of nuclear power, the issue of waste disposal strongly dominates public opinion. In the eyes of the public, the problem of waste disposal is not fully solved, especially in terms of environmental and social acceptability. The public attitude to developing sites for waste disposal has been succinctly summarised in suitable acronyms - NIMBY (not in my back yard), NIMTO (not in my term of office) and BANANA (build absolutely nothing anywhere near anyone)[2]. However, it must also be added that scientific issues related to the disposal problem do indeed need further clarification. Evidence given to the House of Lords Committee[3,4] by the science and engineering community clearly state that "*there is a need for developing the sciences relevant to a detailed assessment, at a fundamental level*".

The source of public attitude seems to be the very long times involving many generations during which this waste must be separated from the biosphere to avoid possible harmful effects. To address this problem, scientists are looking for ways to reduce significantly both the volumes and the radiotoxicity of the waste, and to shorten the very long times for which this waste must be stored safely. For this reason various so-called Partitioning and Transmutation (P&T) techniques are being investigated in most countries with significant nuclear power generating capacity.

Hence there is growing interest in many countries to develop alternative options for the management of spent reactor fuel. These new concepts include long term interim storage and retrievable or non-retrievable final storage of spent fuel and new options for reprocessing.

Apart from the classical aqueous reprocessing of spent fuel, advanced pyro-chemical reprocessing techniques have gained renewed interest. To reduce further the radiotoxicity and the volume of highly active waste, new research programmes have started in several countries to partition and transmute long-lived actinides and fission products in dedicated reactor systems.

3.1. Managment of Nuclear Waste

At the present level of nuclear power production in Europe, approximately 2500 tons of fuel is required each year for 145 reactors in operation in the European Union (EU). This fuel generates about 127 gigawatts of electricity (GWe) thereby providing 36% of the total electricity requirements[5]. Based on an average consumption of electricity in countries of the EU of approximately 6400 kilowatt-hours per capita per year[6] the amount of nuclear fuel required is approximately 14 grams per person per year. Since energy production results in an insignificant loss of mass, it also follows that the waste generated is also approximately 14 grams per person per year[7].

The fresh fuel consists essentially of uranium oxide and heat is generated from the fissioning of uranium atoms in the reactor. In addition to heat generation from fission, various other by-products are generated by nuclear reactions in the reactor. These by-products consist mainly of the element plutonium (Pu), the so-called Minor Actinides (MAs) neptunium, americium, and curium, and fission products (FPs). During reactor operation, the fission products build-up to such a level that they start to "poison" and reduce the mechanical integrity of the fuel. At this stage the *spent* fuel has to be removed from the reactors and be replaced by fresh fuel. Following removal from the reactor, most of the radioactivity in the spent fuel is in the form of the fission products ^{137}Cs, ^{90}Sr and other short-lived nuclides. The radioactivity of these products decreases significantly on time scales of a few decades. Thereafter, the long-lived by-products determine the radioactivity. An estimate of the amounts of these by-products and their half-lives is given in table 2.

For this reason, their disposal requires isolation from the biosphere in a stable deep geological formation for at least tens of thousands of years. Since these products are embedded in the oxide matrix, this means that all 2500 tons per year have to be isolated. If it were possible to separate these products from the spent fuel, and use nuclear reactors or other means to convert them into stable or shorter-lived nuclides, this would greatly facilitate the disposal and management of wastes and raise public acceptance of nuclear energy. This is the motivation behind the Partitioning and Transmutation activities worldwide.

3.2. Partitioning: Advanced Aqueous and Pyro- Processing

If all spent fuel in the EU were to be reprocessed, approximately 2500 tons will have to be treated annually. Current reprocessing technology is based on the aqueous PUREX process in which the spent fuel is dissolved in nitric acid. Using organic solvents, uranium and plutonium are recovered (partitioned) and can be used for fresh fuel fabrication. Thereafter, the aqueous raffinate containing the minor actinides and fission products is vitrified for disposal. The technique can also be extended for the extraction of neptunium, but americium and curium cannot be separated directly in this process.

Table 2. Estimated annual production rates of plutonium, minor actinides and selected fission fragments from all reactors operating in the EU.

Element	Amount (tons per year)	Half-life (years) (most abundant isotope)
plutonium	19	$2.41x10^4$ (^{239}Pu)
minor actinides MA:		
neptunium	2.5	$2.14x10^6$ (^{237}Np)
americium	0.6	433 (^{241}Am)
curium	0.2	18.1 (^{244}Cm)
selected fission fragments:		
technetium-99	2	$2.13x10^5$
iodine-129	1	$1.57x10^7$
caesium-137	3	30.0
strontium-90	1	29.1

Additional advanced aqueous processes have been proposed (TRUEX, DIDPA, TRPO, DIAMEX) in which americium, and curium can be separated. The effectiveness of the separation processes has been investigated at ITU where it has been shown that aqueous methods are technically feasible. Optimisation, however, is required for industrial scale operation.

So-called "dry" (in contrast to "wet" or aqueous) processing techniques may offer a way to avoid problems encountered with aqueous processes. In these techniques the actinides are separated using electrochemical and pyro-chemical separation in molten salts. A key advantage of such processes is that higher levels of radiation can be tolerated. A consequence of this is that reprocessing becomes feasible in which spent fuel that has been cooled for short periods can be processed. In addition, pyro-processing promises compactness, simplicity, low cost (less transport and waste when reprocessing is on-site), and proliferation resistance (due to low product purity). At ITU a pyro-processing laboratory is under construction for such investigations.

Partitioning and Transmutation (P&T) is a scientific and technical challenge involving safety, economic and political issues. Current national and international programmes focus on the following aspects:

- Development and demonstration of aqueous and pyro-chemical processes to separate and recover minor actinides (Np, Am, Cm) and long-lived fission products (^{99}Tc, ^{129}I, ^{135}Cs) from spent fuel/highly active waste
- Development and demonstration of methods for the fabrication of targets and fuel containing minor actinides and long-lived fission products
- Irradiation and post-irradiation experiments to evaluate the performance characteristics of fuel candidates and
- Basic material science studies on inert matrix fuel ("uranium-free" matrix) and other fuel candidates (oxides, metal, nitride, CERMET, CERCER fuels)

Major R+D programmes are underway in China, France, India, Japan, the Russian Federation, U.S., and under European co-operative projects. Important new results have been obtained in the separation efficiencies and recovery rates of minor actinides from highly active wastes.

3.3. Alternative Nuclear Materials - Neptunium and Americium

In 1999 the IAEA Board of Governors issued a recommendation concerning the proliferation risk arising from the minor actinides neptunium (being significantly lower than that of uranium or plutonium) and americium (again being significantly lower than that of neptunium). The IAEA adopted a set of measures designed to provide the international community with the assurance that in the non-nuclear weapon states, available amounts of neptunium or americium in separated form do not raise proliferation concerns. Neptunium monitoring will be implemented on a voluntary basis (regarding the production and transfer of separated neptunium). The monitoring of americium is deferred until a later date. Although little has been separated, neptunium and americium are contained in spent fuel or reprocessing waste. In the European Union, about 4 metric tons are expected to be produced each year in discharged fuel from reactors. Consequently, the envisaged controls are facing basically two challenges: controls on separated quantities of neptunium and verification that neptunium is not clandestinely separated from spent fuel or reprocessing waste.

3.4. Transmutation

Already in the sixties, the transition from Light Water Reactors (LWRs) to Fast Neutron Reactors (FNRs) was foreseen in which plutonium generated in the LWRs would be separated through aqueous processing and used to fabricate fuel for fast reactors. Recent studies have shown that such an approach can be extended to the minor actinides Np, Am, Cm. However, for various technical and political reasons such a transition has not been realised and LWRs still are, and will continue to be, the preferred nuclear power generating reactors.

Accelerator driven systems (ADS)[8] open the possibility of "burning" or incinerating waste material from LWRs in dedicated actinide burners. These actinide burners can burn large quantities of minor actinides (in contrast to critical reactors) safely, and generate heat and electricity in doing so. In addition, schemes have been proposed, in which the long-lived fission products are also destroyed. An advantage of ADS is that, since there is no criticality condition to fulfil, almost any fuel composition can be used in the system. Although these ADS are usually based on the use of fast neutrons, epithermal neutron systems are also being considered.

4. CASE STUDY: ACCELERATOR DRIVEN SYSTEMS (ADS)

In the present section we describe how a potential proliferation problem was identified in the relatively new and emerging field of accelerator driven systems (ADS) and how legislation has been introduced recently to cover misuse of this technology.

Over the past few years we have witnessed a surge of interest in the use of accelerator technology for nuclear applications. In particular, accelerator driven systems for nuclear power production and nuclear waste burning have been proposed. Such systems may indeed play a role in Generation IV nuclear systems. The DOE in the U.S. has recently issued in 1999 a "roadmap" for the accelerator transmutation of waste. In Europe, eight countries and the Joint Research Centre of the European Commission have formed a technical working group on ADS to develop a roadmap for "Nuclear Waste Transmutation using Accelerator Driven Systems". This report should be finalised in early 2001. Similar activities are underway in Japan and Korea.

The general approach in ADS is fairly similar. The various schemes based on a double strata concept are outlined in fig. 1.

The first stratum is based on a conventional fuel cycle and consists of standard LWRs and FNRs, fuel fabrication and reprocessing plants. The recovered plutonium is recycled as mixed oxide fuel in the thermal and fast reactors. The remaining plutonium, MAs and long-lived fission products (Japan only) are partitioned from the waste and enter the second stratum where they are transmuted in a dedicated ADS.

In the second stratum, devoted primarily to waste reduction, the Pu, MAs, and long-lived fission products (Japanese concept) are fabricated into fuels and targets for transmutation in dedicated ADS. The use of dry reprocessing in this stratum allows for multiple reprocessing of the fuel. A key advantage of this is that higher levels of radiation can be tolerated in the molten salts and allows reprocessing of spent fuel which has been cooled for periods as short as one month. In the "*Energy Amplifier*" concept proposed by CERN, the objective is to burn actinides and fission products within the thorium fuel cycle. The advantage of using thorium lies in the fact that less transuranium elements are produced. In the Minimum Scheme, waste from a LWR is processed and some of the separated plutonium is recycled into the thermal reactor. The remaining plutonium, the MAs and FPs together with thorium are used to fabricate fuel for the ADS waste burner. In the U.S., P&T is foreseen for treatment of the spent fuel arising from a once-through fuel cycle. Hence, rather than being an integral part of the fuel management scheme, the spent fuel is processed (using a UREX process) to initially separate the uranium, stable and short lived FPs, which are disposed of as low level waste. The remaining Pu, MAs, and FPs are then sent for transmutation in the ADS or ATW system (accelerator transmutation of waste).

In contract to standard *critical* nuclear reactors in which there are enough neutrons to sustain a chain reaction, *sub-critical* systems used in ADS need an external source of neutrons to sustain the chain reaction. These "extra" neutrons are provided by the accelerator. More exactly the accelerator produces high-energy protons which then interact with a spallation sources to produce neutrons.

But why go to all this trouble to build an ADS sub-critical system when critical reactors systems already work? The answer to this lies in the fact that one has more control and flexibility in the design of the sub-critical reactor. This is required when the reactor is being used to transmute large amounts of nuclear waste in the form of minor actinides (MAs). Today it appears that ADS has great potential for waste transmutation and that such system may go a long way in reducing the amounts of waste and thereby reducing the burden to underground repositories.

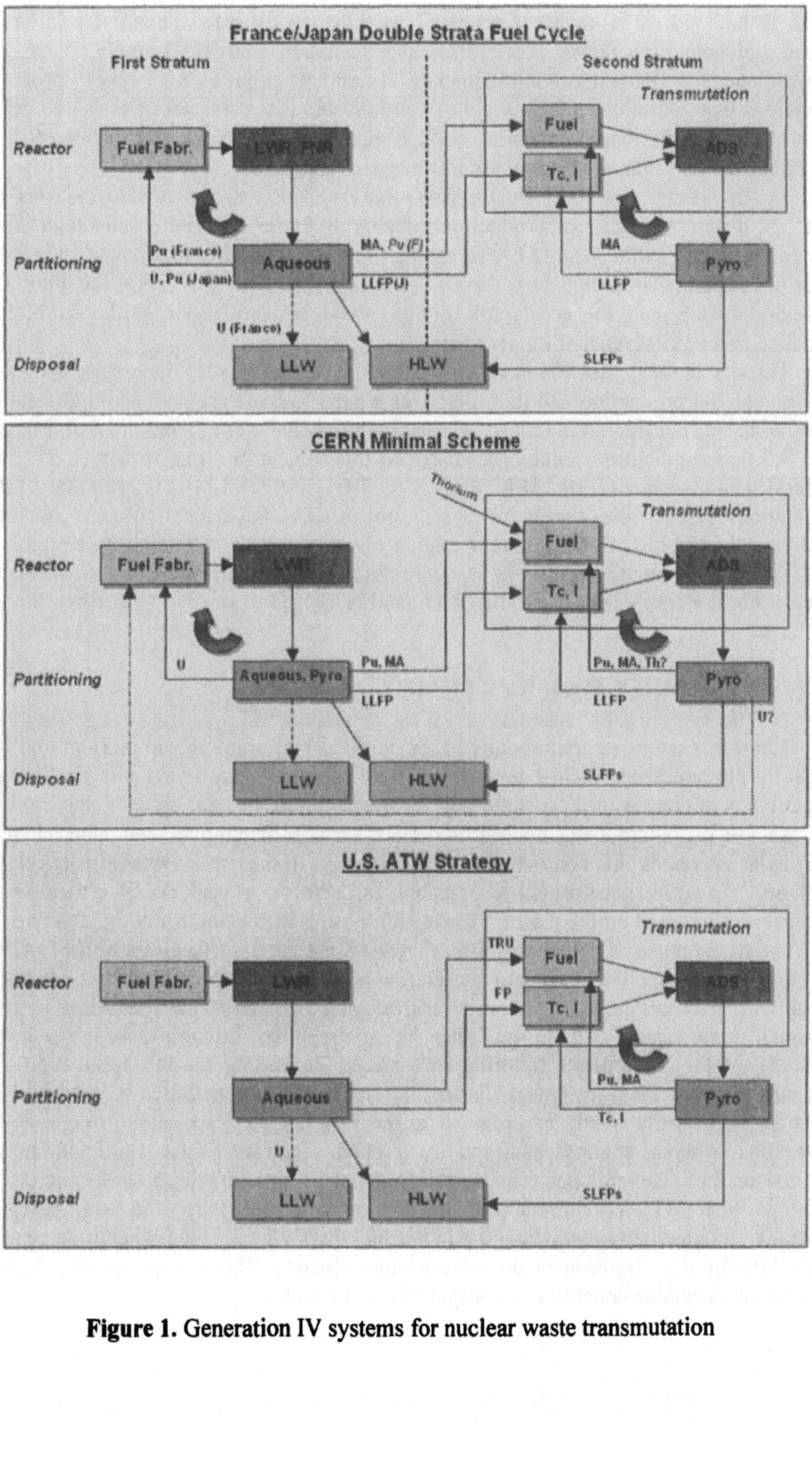

Figure 1. Generation IV systems for nuclear waste transmutation

At ITU, we have been involved in assessing the proliferation risks associated with ADS technology. In a series of articles[9], we have investigated potential proliferation aspects of both high power accelerators and spallation sources. Actually, it has been known since the 1940s that bombardment of a uranium target by high-energy protons or deuterons would produce a large yield of neutrons and that these neutrons can be used in turn to produce fissionable material through nuclear reactions. In 1941, Glenn Seaborg produced the first man-made plutonium using an accelerator.

In our articles, we have shown that relatively small commercial cyclotrons (150 MeV, 2 mA) are capable of producing amounts of fissile material greater than the so-called "screening limit". The IAEA screening limit is a capability to produce 100g Pu per year or to operate continuously at thermal powers greater than 3 MW. On the other hand, accelerators foreseen for accelerator driven waste transmutation can be expected to produce tens of kilograms of Pu per year.

Through these studies, we have shown that this technology is now mature enough for fissile material production and that there was a need to issue regulations on this subject. This work was considered in issuing of recent regulations[10]. The Department of Energy in the U.S. has announced regulation changes on this subject in a recent (March 27, 2000) federal announcement (10 CFR Part 810, RIN 1992-AA24). In summary, these regulations restrict the export of accelerator driven sub-critical systems and their components capable of continuous operation above 5 MWth. Although accelerators are not specifically mentioned, there is some room for interpretation as to whether an accelerator is a component of a "utilization facility".

5. EXOTIC NUCLEAR SYSTEMS

Nuclear reactors are traditionally based on the use of uranium and thorium and more recently plutonium as nuclear materials. The current safeguards systems are designed specifically to cope with these materials. However, future nuclear systems may be based on the use of "alternative" nuclear materials such as neptunium and americium. Of particular interest in this context are recent proposals to use americium (in particular the isotope ^{242m}Am) in space propulsion systems and neutron amplifiers[11]. A critical or near critical system based on the use of thin actinide films is in schematically fig. 2.

The fissile material constitutes a thin layer on the inner surface of a hollow cylinder of circular cross-section, made of a neutron moderator material such as graphite or beryllium. Such systems can be made critical with relatively small amounts of fissile material. In the case of ^{242m}Am, less than 100 g is required. The thickness of the layer is typically in the micrometer to millimetre range, depending on the application. This thickness depends upon the type of fissile material and its concentration in this layer - but must be sufficiently small in order to allow fast neutrons to pass through without interaction, whereas thermal neutrons are trapped. Neutrons in the Hohlraum may be thermal or fast. Thermal neutrons react immediately with the layer and generate fast neutrons whereas fast neutrons pass through without interaction. In both cases fast neutrons penetrate into the moderator and become thermalized. If these neutrons penetrate again into the thin fissile layer they cause more fissions. Those which escape from the cylinder at its outside constitute the output of the assembly.

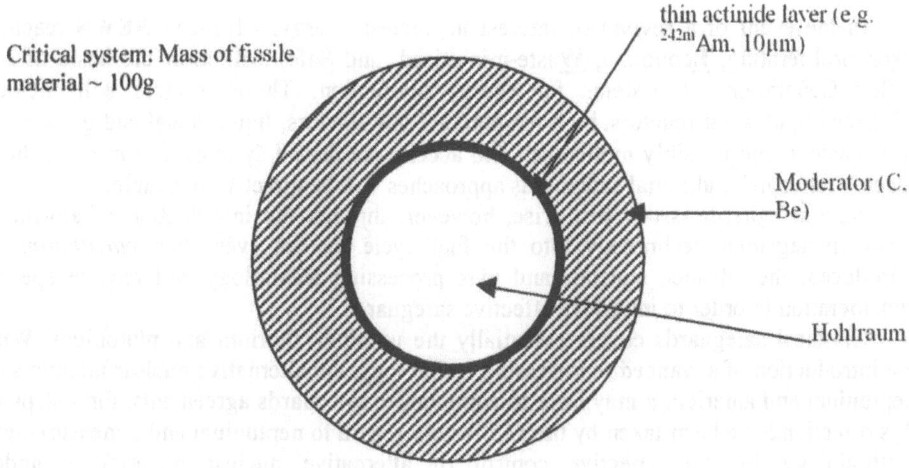

Figure 2. A critical or near critical system based on a Hohlraum, a thin actinide layer and a moderator

Is the use of the isotope ^{242m}Am in such systems a proliferation issue? It is straightforward to show, from the decay scheme of ^{242m}Am, that even relatively small amounts give rise to considerable radiation in the form of gamma photons, neutrons emission, and heat production. On this basis, we believe that in itself ^{242m}Am is not a proliferation issue.

6. THE COST OF SAFEGUARDS

In the previous sections we have described non-proliferation and safeguards aspects of candidate nuclear reactors for Generation IV power systems and their associated technology. Since one of the key criteria for these future reactors systems is economic competitivity, we must ask the question: What does safeguards cost?

In the 1999 annual report of the IAEA, the safeguards budget was given as 78 M$. An extrabudgetry figure (mainly for R&D tasks) amounted to 20M$. The total nuclear energy production worldwide in 1999 was 2403 TWh. On this basis, the costs of safeguards can be calculated as $98 \times 10^6 / 2403 \times 10^9 = 4 \times 10^{-5}$ $/kWh or $\underline{4 \times 10^{-3}}$ cents/kWh. This is negligible in comparison to the overall costs of electricity of 3 cents/kWh based on year 2000.

Similar figures are quoted by EURATOM. The total safeguards costs for 1999 'were 45 M€ (maintenance & running costs 12M€, staff credits 27M€, mission costs 6 M€). The income side of Euratom is not structured, it comes out of the commission budget which is fed by the member states VAT and import (into the EU) taxes as well as other additional small sources. The total European nuclear energy amounted to 825 TWh in 1999. The resultings safeguards costs can be given as $45 \times 10^6 / 825 \times 10^9 =$ $\underline{5.5 \times 10^{-5}}$ €/kWh

7. CONCLUSIONS

In the event of a revival of interest in nuclear energy, advanced **NEWS** reactors (Non-proliferating, Economic, Waste-minimised, and Safe) will form the basis of so-called Generation IV systems for energy production. These reactors will include advanced light water reactors, high temperature gas reactors, liquid metal and gas-cooled fast reactors, and possibly molten salt and accelerator driven systems. For most of these reactor concepts, traditional safeguards approaches will be directly applicable.

New safeguards issues will arise, however, through the introduction of advanced waste management technology into the fuel cycle. In the event that *partitioning* is introduced, the advance aqueous and pyro-processing technology will require special consideration in order to introduce effective safeguards.

Classical safeguards covers essentially the uranium, thorium and plutonium. With the introduction of advanced reprocessing technology, the alternative nuclear materials of neptunium and americium may have to come under safeguards agreements. First steps in this direction have been taken by the IAEA with regard to neptunium and a measurement methodology for the effective control of alternative nuclear material is under development. However it should be added that no purified and directly usable end products are supplied by these type of facilities. The materials are always handled and processed remotely in high radiation fields, including their recycling in burner reactors. At research facilities, however, there is a dedicated effort to develop and test optimum processes for the recovery and characterisation of minor actinides (neptunium, americium, curium). They are usually separated or handled in pure forms in order to determine decontamination factors and recovery rates. The amounts of material which are used are however small in comparison to the significant quantities of these materials.

In the event that *transmutation* reactor systems are introduced, special consideration must be given to the technology. In particular potential proliferation problems associated with accelerator driven systems (ADS) for waste transmutation have already been identified. Initial studies, however, have shown how such reactor systems can be safeguarded. In addition, new regulations covering the export of related technology have been introduced recently in the U.S.

Exotic nuclear systems may also play a part in Generation IV systems. They are likely to be compact and contain relatively small amounts of fissile material. Examples of systems based on thin layers of fissile material have been described. These may find application in space propulsion reactors or as boosted spallation sources. Although they do require further investigation, it is not expected that they will present a problem from the safeguards viewpoint.

Finally, it has also shown that the costs associated with safeguarding present day civilian reactors are very small indeed compared to the overall capital investment and operational costs, and that safeguarding Generation IV systems will not have an overall negative impact with regard to non-proliferation.

8. REFERENCES

1. Generation IV Nuclear Energy Systems Initiative, http://gen-iv.ne.doe.gov/
2. D. R. Williams, *What is Safe? The Risks of Living in a Nuclear Age*, The Royal Society of Chemistry, Cambridge, 1998
3. The Royal Society, *Nuclear Energy, The Future Climate*, p27, 1999
4. *Management of nuclear wastes*, Volume II (Evidence) (Select Committee on Science and Technology House of Lords, para 925, p146, March 1999.)
5. 1999 World Nuclear Industry Handbook, Nuclear Engineering International, p12. (European Union countries with nuclear power are Belgium, Finland, France, Germany, Netherlands, Spain, Sweden, and UK). Assuming a conversion efficiency of 35% thermal to electrical energy.
6. Informations Utiles, 1999, p. 16, CEA, ISBN 2-11-091547-1.
7. L. Koch, J.-P. Glatz, R. J. M Konings, J. Magill, Partitioning and Transmutation Studies at ITU, in Institute for Transuranium Elements Annual Repirt, 1999. See under Publications: http://itu.jrc.cec.eu.int .
8. J. Magill, P Peerani, J. van Geel, Closing the Fuel Cycle with ADS, International Workshop on the Physics of Accelerator-Driven Systems for Nuclear Transmutation and Clean Energy Production, 29th Sept. - 3rd Oct. 1997, Trento, Italy. http://itumagill.fzk.de/ADS/trento/trento.html.

   M. Salvatores, Studies at CEA-France on the Role of Accelerator-Driven systems in Waste Incineration Scenarios, Proceedings of the I.A.E.A. Technical Committee Meeting on: Feasibility and Motivation for Hybrid Concepts for Nuclear Energy Generation and Transmutation, Madrid, Spain, 17-19 September 1997, to be published.

   T. Mukaiyama, et al., Omega Program & Neutron Science Project for Development of Accelerator Hybrid System at JAERI, Proceedings of the I.A.E.A. Technical Committee Meeting on: Feasibility and Motivation for Hybrid Concepts for Nuclear Energy Generation and Transmutation, Madrid, Spain, 17-19 September 1997, to be published.

   C Rubbia, CERN Concept of ADS, Proceedings of the I.A.E.A. Technical Committee Meeting on: Feasibility and Motivation for Hybrid Concepts for Nuclear Energy Generation and Transmutation, Madrid, Spain, 17-19 September 1997, to be published.
9. J. Magill, P. Peerani, R. Schenkel, J. van Geel, Accelerators and (Non-) Proliferation, Proceedings of the Topical Meeting in Nuclear Applications of Accelerator Technology, Nov. 16-20, 1997, Albuquerque, New Mexiko, ANS,. p440-446.

   J. Magill, P. Peerani, Proliferation Issues. In: Thorium as a Waste Management Option, Chapt. 2.6 (1999) 117-131, Nuclear Science and Technology, European Commission, Report EUR 19142EN.

   J. Magill, P. Peerani, (Non-) Proliferation Aspects of Accelerator Driven Systems, Journal de Physique IV, Vol. 9 (1999) p167-181. http://itumagill.fzk.de/ADS/LesHouches/LesHouches.pdf.

   J. Magill, R. Schenkel, L. Koch, Verification of Nuclear Materials and Sites, Detection of Clandestine Activities, and Assessment of the Proliferation Resistance of New Reactors and Fuel Cycle Concepts, Proceedings of the Fachsitzung zum Thema "Abrüstung und Verification" Dresden 22-23 März, 2000, to be published.

   J. Magill, P. Peerani, Proliferation Aspects of ADS. In: Impact of accelerator-based technologies on nuclear fission safety - IABAT project, European Commission, Nuclear Science and Technology, report EUR 19608 EN, p60-67, 2000.

   R. Schenkel, M. Betti, K. Mayer, J. Magill, New Safeguards Issues with Advanced Nuclear Technologies, Proceedings of the INMM/ESARDA 3rd Workshop on "Science and Modern Technology for Safeguards" 13-16 Nov. 2000, Tokyo, to be published.
10. Office of Defense Nuclear Nonproliferation, U.S. DOE, Assistance to Foreign Atomic Energy Activities, Federal Register, vol. 65 no. 59, March 272000 (see http://itumagill.fzk.de/ADS/getdoc.pdf).
11. J. Magill, P. Peerani, and J. van Geel, *A Neutron Amplifier Assembly, European* Patent Application No. 99107327.1, Fo 2619 EP, 20 April 1999, Munich.

    J. Magill, P. Peerani, and J. van Geel, Basic Aspects of Sub-Critical Systems based on Thin Actinide Films, 3. International Conference on Accelerator Driven Transmutation Technologies and Applications June 7-11, 1999 Prague (Czechoslovakia).

    Y. Ronen, M. Aboudy, D. Regev, A nuclear engine design 242mAm as a nuclear fuel, Annals of Nuclear Energy, 27 (2000) 85-91.

    C. Rubbia, Neutrons in a highly diffusive, transparent medium: an effective neutron 'storage' device. Trans. ICENES '98, the Ninth International Conference on Emerging Nuclear Energy Systems, Tel-Aviv, Israel, June 28 - July 2, 1998, Vol.1, page 4. Dan Knassin Ltd. Ramat-Gan, P.O.Box 1931, Israel.

BLACKLIGHT POWER TECHNOLOGY
A New Clean Energy Source with the Potential for Direct Conversion to Electricity

Randell L. Mills*

1. INTRODUCTION

BlackLight Power, Inc. (the Company), a Delaware corporation based in its 53,000 sq. ft. headquarters in Cranbury, New Jersey, believes it has developed a new hydrogen chemical process that generates power, plasma (a hot ionized glowing gas), and a vast class of new compositions of matter. Specifically, the Company has designed and tested a new proprietary energy-producing chemical process. The Company has developed high-power density, high-temperature, hydrogen gas cells that produce intense light, power of orders of magnitude greater than that of the combustion of hydrogen at high temperatures, and power densities equal to those of many electric power plants. The Company is focusing on cells for generating light and plasma for lighting applications and direct conversion to electricity, respectively.

The cells generate energy through a chemical process (BlackLight Process) which the Company believes causes the electrons of hydrogen atoms to drop to lower orbits thus releasing energy in excess of the energy required to start the process. The lower-energy atomic hydrogen product of the BlackLight Process reacts with an electron to form a hydride ion which further reacts with elements other than hydrogen to form novel compounds called hydrino hydride compounds (HHCs) which are proprietary to the Company. The Company is developing the vast class of proprietary chemical compounds formed via the BlackLight Process. Its technology has far-reaching applications in many industries.

The power may be in the form of a plasma, a hot ionized glowing gas. The plasma may be converted directly to electricity with high efficiency using a known microwave device called a gyrotron, thus, avoiding a heat engine such as a turbine. The Company is

* Randell L Mills, President, BlackLight Power, Inc., 493 Old Trenton Road, Cranbury, NJ 08512, Phone: 609-490-1090, e-mail: rmills@blacklightpower.com; www.blacklightpower.com

Global Warming and Energy Policy, Edited by Kursunoglu *et al.*
Kluwer Academic/Plenum Publishers, New York 2001

working on direct plasma to electricity conversion. The device is linearly scaleable from the size of hand held units to large units which could replace large turbines.

There are many advantages of the technology. The energy balance permits the use of electrolysis of water to split water into its elemental constituents of hydrogen and oxygen as the source of hydrogen fuel using a small fraction of the output electricity. Additionally, pollution produced by fossil and nuclear fuels should be eliminated since no green house gases, air pollutants, or hazardous wastes are produced. As no fossil fuels are required, the projected commercial operating costs are much less than that of any known competing energy source.

The Company's process may start with water as the hydrogen source and convert it to HHCs; whereas, fuel cells typically require a hydrocarbon fuel and an expensive reformer to convert hydrocarbons to hydrogen and carbon dioxide. The Company's plasma to electric conversion technology with no reformer, no fuel cost, creation of a valuable chemical by-product rather than pollutants such as carbon dioxide, and significantly lower capital costs and operating and maintenance (O&M) costs are anticipated to result in household units that are competitive with central power and significantly superior to competing microdistributed power technologies such as fuel cells.

2. THE BLACKLIGHT PROCESS

Based on physical laws of nature, Dr. Mills theory predicts that additional lower energy states are possible for the hydrogen atom, but are not normally achieved because transitions to these states are not directly associated with the emission of radiation. Thus, the ordinary hydrogen atom as well as lower-energy hydrogen atoms (termed hydrinos by Dr. Mills) are stable in isolation. Mills theory further predicts that hydrogen atoms can achieve these states by a radiationless energy transfer with a nearby atom, ion, or combination of ions (a catalyst) having the capability to absorb the energy required to effect the transition. Radiationless energy transfer is common. For example, it is the basis of the performance of the most common phosphor used in fluorescent lighting. Thus, the Company believes hydrogen atoms can be induced to collapse to a lower-energy state, with release of the net energy difference between states. Successive stages of collapse of the hydrogen atom are predicted, resulting in the release of energy in amounts many times greater than the energy released by the combustion of hydrogen. Since the combustion energy is equivalent to the energy required to liberate hydrogen from water, a process which takes water as a feed material and produces net energy is possible. The equivalent energy content of water would thus be several hundred to several thousand times that of crude oil, depending on the average number of stages of collapse.

The Company is the pioneer of technology based on the chemical process of releasing further chemical energy from hydrogen called the "BlackLight Process." More specifically, energy is released as the electrons of hydrogen atoms are induced by a catalyst to transition to lower-energy levels (i.e. drop to lower base orbits around each atom's nucleus) corresponding to fractional quantum numbers. The lower energy atomic hydrogen product is called "hydrino," and the hydrogen catalyst to form hydrino is called a "transition catalyst." As hydrogen atoms are normally found bound together as molecules, a hot dissociator is used to break hydrogen molecules into individual hydrogen atoms. A vaporized catalyst then causes the normal hydrogen atoms to transition to lower-energy states by allowing their electrons to fall to smaller radii around the nucleus with a release of energy that is intermediate between chemical and nuclear energies. The products are power, plasma, light, and novel HHCs.

The catalysts used and the BlackLight Process are the proprietary intellectual property of the Company. The theory, data, and analysis supporting the existence of this new form of energy have been made publicly available.[1-15] Also see the BlackLight Power web page: www.blacklightpower.com. Laboratory scale devices demonstrating means of extracting the energy have been operated at the Company and at independent laboratories. Results to date indicate that the process can eventually provide economically competitive products in a wide range of applications including lighting, thermal, and electric power generation. The Company's gas energy cells, even in prototype stage, are frequently operating at power densities and temperatures equivalent to those of many coal fired electric power plants and produce about 100 times the energy of the combustion of the hydrogen fuel. The plasma is permissive of a direct plasma to electricity conversion technology as well as the production of electricity by conventional heat engines. The Company currently believes that the scale-up of energy cells to commercial power generation level will require mainly the application of existing industry knowledge in catalysis and power engineering.

The lower-energy atomic hydrogen product of the BlackLight Process reacts with an electron to form a hydride ion which further reacts with elements other than hydrogen to form novel compounds which are proprietary to the Company. The Company is developing the vast class of proprietary chemical compounds formed via the BlackLight Process. Test results indicate that the properties of HHCs are rich in diversity due to their extraordinary binding energy (i.e. the energy required to remove an electron which determines the chemical reactivity and properties). This new class of matter may be comparable to carbon in terms of the possibilities of new compositions of matter. Carbon is a base element for many useful compounds ranging from diamonds, to synthetic fibers, to liquid gasoline, to pharmaceuticals. The Company believes hydrino hydride ions have the potential to be as useful as carbon as a base "element." The novel compositions of matter and associated technologies have far-reaching applications in many industries including the chemical, lighting, computer, defense, energy, battery, propellant, munitions, surface coatings, electronics, telecommunications, aerospace, and automotive industries. The Company is focusing on developing a high voltage battery and silane materials based on the novel hydride chemical products. Many additional applications of the chemical compounds are possible.

2.1 Validation

Based on the solution of a Schrödinger-type wave equation with a nonradiative boundary condition based on Maxwell's equations, Mills[1-15] predicts that atomic hydrogen may undergo a catalytic reaction with certain atomized elements or certain gaseous ions which singly or multiply ionize at integer multiples of the potential energy of atomic hydrogen, 27.2 eV.

For example, potassium ions can provide a net enthalpy of a multiple of that of the potential energy of the hydrogen atom. The second ionization energy of potassium is 31.63 eV, and K^+ releases 4.34 eV when it is reduced to K.[16] The combination of reactions K^+ to K^{2+} and K^+ to K, then, has a net enthalpy of reaction of 27.28 eV.

The reaction involves a nonradiative energy transfer to form a hydrogen atom that is lower in energy than unreacted atomic hydrogen. The product hydrogen atom has an energy state that corresponds to a fractional principal quantum number. Recent analysis of mobility and spectroscopy data of individual electrons in liquid helium show direct experimental confirmation that electrons may have fractional principal quantum energy levels.[14] The lower-energy hydrogen atom is a highly reactive intermediate which further reacts to form a novel hydride ion. Single line emission from the excited catalyst ion having accepted the energy from atomic hydrogen has been observed with the emission from the predicted hydride ion product.[15] Thus, the catalytic reaction with the formation the novel hydride ions is confirmed spectroscopically.

Typically, the emission of extreme ultraviolet (EUV) light from hydrogen gas is achieved by a discharge at high voltage, a high power inductively coupled plasma, or in hot fusion research, a plasma is created and heated by radio waves to 10s of millions of degrees with confinement of the hot plasma by a toroidal (donut shaped) magnetic field. The Company has observed intense EUV emission at low temperatures from atomic hydrogen and certain atomized pure elements or certain gaseous ions which ionize at integer multiples of the potential energy of atomic hydrogen. The Company has tested over 130 elements and compounds which covers essentially all of the elements of the periodic chart. The chemical interaction of catalysts with atomic hydrogen at temperatures below 1000 K has shown surprising results in terms of the emission of the Lyman and Balmer lines[2-7] (atomic hydrogen emission ten times more energetic than the combustion of hydrogen), emission of lines corresponding to lower-energy hydrogen states and the corresponding hydride ions, and the formation of novel chemical compounds.[8-13]

Over 20 independent labs have performed 25 types of analytical experiments that confirm the Company's novel catalytic reaction of atomic hydrogen which produces an anomalous discharge or plasma and produces novel hydride compounds.[2-13]

Experiments that confirm the novel hydrogen chemistry include extreme ultraviolet (EUV) spectroscopy, plasma formation, power generation, and analysis of chemical compounds. For example:

1. Pennsylvania State University Chemical Engineering Department has determined heat production associated with hydrino formation with a Calvet calorimeter that showed the generation of 10^7 *J/mole* of hydrogen, as compared to $2.5X10^5$ *J/mole* of hydrogen anticipated for standard hydrogen combustion.[17] Thus, the total heats generated appear to be 100 times too large to be explained by conventional chemistry, but the results are completely consistent with Mills model.

2. Lines observed by EUV spectroscopy could be assigned to transitions of atomic hydrogen to lower-energy levels corresponding to lower-energy hydrogen atoms and to the emission from the excitation of the corresponding hydride ions.[4]

For example, the product of the catalysis of atomic hydrogen with potassium metal, $H[a_H/4]$ may serve as both a catalyst and a reactant to form $H[a_H/3]$ and $H[a_H/6]$. The transition of $H[a_H/4]$ to $H[a_H/6]$ induced by a multipole resonance transfer of 54.4 *eV* (2·27.2 *eV*) and a transfer of 40.8 *eV* with a resonance state of $H[a_H/3]$ excited in $H[a_H/4]$ is represented by $H[a_H/4] + H[a_H/4] \rightarrow H[a_H/6] + H[a_H/3] + 176.8\ eV$.

The predicted 176.8 *eV* (70.2 Å) photon is a close match with the 73.0 Å line observed by a team headed by Dr. Johannes P. F. Conrads, then Director and Chairman of the Board, of Institut Fur Niedertemperatur–Plasmaphysik e.V. and the Ernst–Moritz Arndt–Univeristat Greifswald ("INP"), a top plasma physics laboratory in Greifswald, Germany. The energy of this line emission corresponds to an equivalent temperature of 1,000,000 °C and an energy over 100 times the energy of combustion of hydrogen.

3. Observation of intense extreme ultraviolet (EUV) emission has been reported at low temperatures (e.g. $\approx 10^3$ K) from atomic hydrogen and certain atomized elements or certain gaseous ions.[2-7] The only pure elements that were observed to emit EUV were those wherein the ionization of t electrons from an atom to a continuum energy level is such that the sum of the ionization energies of the t electrons is approximately $m \cdot 27.2\ eV$ where t and m are each an integer. Potassium, cesium, and strontium atoms and Rb^+ ion ionize at integer multiples of the potential energy of atomic hydrogen and caused emission. Whereas, the chemically similar atoms, sodium, magnesium, and barium do not ionize at integer multiples of the potential energy of atomic hydrogen and caused no emission.

4. An energetic plasma in hydrogen was generated using strontium atoms as the catalyst. The plasma formed at 1% of the theoretical or prior known voltage requirement with 4,000–7,000 times less power input power compared to noncatalyst controls, sodium, magnesium, or barium atoms, wherein the plasma reaction was controlled with a weak electric field.[2, 7]

5. An anomalous plasma with hydrogen/potassium mixtures has been reported wherein the plasma decayed with a two second half-life which was the thermal decay time of the filament which dissociated molecular hydrogen to atomic hydrogen when the electric field was set to zero.[5-6] This experiment showed that hydrogen line emission was occurring even though the voltage between the heater wires was set to and measured to be zero and indicated that the emission was due to a reaction of potassium atoms with atomic hydrogen which confirms a new chemical source of power.

Reports of the formation of novel compounds provide substantial evidence supporting a novel reaction of hydrogen as the mechanism of the observed EUV emission and anomalous discharge. Novel hydrogen compounds have been isolated as products of the reaction of atomic hydrogen with atoms and ions identified as catalysts in the reported EUV studies.[2-13] Novel inorganic alkali and alkaline earth hydrides of the formula MH^* and MH^*X wherein M is the metal, X, is a singly negatively charged anion, and H^* comprises a novel high binding energy hydride ion were synthesized in a high temperature gas cell by reaction of atomic hydrogen with a catalyst such as potassium metal and MH, MX or MX_2, corresponding to an alkali metal or alkaline earth metal compound, respectively.[8, 11] Novel hydride compounds were identified by (1) time of flight secondary ion mass spectroscopy which showed a dominate hydride ion in the negative ion spectrum, (2) X-ray photoelectron spectroscopy which showed novel hydride peaks and significant shifts of the core levels of the primary elements bound to the novel hydride ions, (3) proton nuclear magnetic resonance spectroscopy (NMR) which showed extraordinary upfield chemical shifts compared to the NMR of the corresponding ordinary hydrides, and (4) thermal decomposition with analysis by gas chromatography, and mass spectroscopy which identified the compounds as hydrides.[8,11]

An upfield shifted NMR peak is consistent with a hydride ion with a smaller radius as compared with ordinary hydride since a smaller radius increases the shielding or diamagnetism. Thus, the NMR shows that the hydride formed in the catalytic reaction has been reduced in distance to the nucleus indicating that the electrons are in a lower-energy state. Compared to the shift of known corresponding hydrides the NMR provides direct evidence of reduced energy state hydride ions.

The NMR results confirm the identification of novel hydride compounds MH^*X, MH^*, and MH_2^* wherein M is the metal, H, is a halide, and H^* comprises a novel high binding energy hydride ion. For example, large distinct upfield resonances were observed at –4.6 ppm and –2.8 ppm in the case of KH^*Cl and KH^*, respectively. Whereas, the resonances for the ordinary hydride ion of *KH* were observed at 0.8 and 1.1 ppm. The presence of a halide in each compound *MH*X* does not explain the upfield shifted NMR peak since the same NMR spectrum was observed for an equimolar mixture of the pure hydride and the corresponding alkali halide (*MH/MX*) as was observed for the pure hydride, *MH*. The synthesis of novel hydrides such as *KH** with upfield shifted peaks prove that the hydride ion is different from the hydride ion of the corresponding known compound of the same composition. The reproducibility of the syntheses and the results from independent laboratories confirm the formation of novel hydride ions.

3. BUSINESS UNITS

The Company believes that it has created a commercially competitive new source of energy, a new source of plasma which releases rather than consumes energy, a new source of light, and a revolutionary new field of hydrogen chemistry. With its achievements of a sustained 100,000+ °C plasma of hydrogen with essentially no power input to its power cell and synthesis of over 40 novel compounds in bulk with extraordinary properties the Company is focusing on product development. Initial target products are a direct plasma to electric power cell targeted at the residential and commercial microdistributed markets and the premium power market. Additional market objectives for the plasma and chemistry technologies are lighting sources, a high voltage battery to power portable electronics and electric vehicles, and chemical products and processes based on silicon and hydrino chemistry.

The Company has two basic business units—power and chemical. The plasma-electric technology may represent a near-term huge energy market. But, in the case of a large central power plant, the Company estimates that the potential revenues from the chemicals produced with power generation may eclipse the electricity sales. However, both offer extraordinary potential revenue and profit. Since enormous power (easily convertible to electricity) is a product of the BlackLight Process, the two units can operate in tandem seamlessly.

The priorities of the Company's power business is the residential and commercial microdistributed markets and the premium power market based on its plasma-electric power cell technology. The time to market should be near term for these relatively small-scale, simple devices that are projected to be inexpensive to manufacture, service, and use, and vastly superior to competing technologies such as internal combustion engine gensets, fuel cells, and microturbines. Selected statistics on electric generation are given in Table 1.

Early adopters of BlackLight power systems are expected to be those that require premium power generated on-site. The premium power market* includes businesses where brief electrical outages can cause severe monetary loss: telecommunications sites, computer centers, server hotels, e-commerce centers, semiconductor fabrication facilities, and others. The market size was estimated to be 30,000 MW in 1999 and growth to be multiples of the entire energy market rate.[18-19] The Glider Group and Stephens Inc.

---

*The premium power market is also known as the 9's market and the powercosm market. Utility grids provide 99.9% reliability, or 8 hours of disruption per year. For the Internet economy even small fractions of a second can cost millions of dollars. In high technology manufacturing industries even hours of disruption can shut down operations for days, again costing millions. More reliability is measured in %, the more 9's required (99.999...%), the smaller the fractions of a second power is disrupted, and the more valuable the power.

Table 1. Statistics on electric generation.

US Electric Market
• $217 billion in annual US sales (1998).
• 43% Residential
• 32% Commercial
• 22% Industrial
• 5% Other
Capital Expenditures Required to Meet New Generation Demand
• Estimated at $90 Billion Globally with 10% in US in 1999
• $21 Billion will be spent on Premium Power in 2000
• $30 Billion in 2002
• $50 Billion in 2005
Premium Power Consumption/Demand
• Estimated to be 30,000 MW in 1999
• Estimated to be 500,000 MW in 2000
• Double digit growth expected over next five years

Table 2. Competitive Advantages of The BlackLight Power Process.

Cost Per KWH of Alternative Energy Sources	
Coal	4-5 ¢
Natural Gas	4-5 ¢
Oil	4-5 ¢
Nuclear Power	5-6 ¢
Hydroelectric	4-7 ¢
Geothermal	5-8 ¢
Wind	5-9 ¢
Solar	10-12 ¢
Photovoltaic	30-40 ¢
BlackLight *	<1 ¢

*Cost figures include operating, maintenance, capital generating expense of plasma-electric system (Source: EPRI, BLP)

estimates[20-21] that this market is 15% of the current US energy market; and will be 30–50% within 3–5 years as the internet economy build-out continues. This market is characterized by early adoption of emerging technologies and an insensitivity to cost. For example, a typical rate is over $1,000 per kWh rate and the rate for the upper-end of the reliability scale, six 9's reliable power, is about $1 million per kWh compared to 5 ¢ per kWh for three 9's power supplied by the grid. The premium power market is a multi-billion dollar market. The current equipment market is $21 billion in hardware alone and is projected to ellipse the profitability of the entire utility market in the near term.[22]

BlackLight's Energy Systems design advantages are: virtually instantaneous turn on/off, simplicity, easy logistics, low capital cost, low operational and maintenance cost, easy redundancy (for reliability), and no pollution. With our current design, BlackLight projects capital costs around $25–100 per kW, and very low generation cost (<$0.01 per kWh). This is lower than competitive solutions, but in this market segment cost is not a driver. Our chief competitors are reciprocating engine-based gensets built by Catapillar, Cummins, and others. Additional competition might be from newer entrants: microturbines and fuel cells. The former competitors, fossil-fueled engines, have an advantage because they are an incumbent technology, but they will not be able to significantly improve their reliability, have a short lifetime, do not meet pollution requirements, and can not reduce their O&M costs to be competitive with our solution. The latter competitors have a slight advantage in name recognition relative to BlackLight, but microturbines and fuel cells are not suited for the premium power market. Fuel cells and turbine systems take too long to start up, and are difficult to harmonize with grid-supplied power. Thus they are ineffective at improving power reliability.

Due to superior performance of its technology, the Company expects early adoption by the premium power market with expansion into the broader microdistributed market. The broader market which includes hundreds of millions of homes and businesses in the US and Europe will be drawn by significant cost savings and increasing unreliability of the grid with a lack of viable microdistributed alternatives. The populace of the third world, particularly Asia, represents a further enormous market opportunity for which

BlackLight technology is particularly suited, since in addition to very low capital and O&M costs, no fuel or electrical grid infrastructure is required.

In terms of its development strategy for large scale systems, the Company has decided to focus on developing the chemical business unit as a first priority over large power plants. In addition to the possibility of larger revenue, the chemical business offers several other initial advantages. A power generation plant based on thermal energy would have to be scaled-up while maintaining current or higher levels of power density before it could be commercialized. Scaling up to a power plant of very large proportions has engineering risks. While there are engineering risks associated with the scale-up for chemical production, they are not as daunting. Some potential product areas such as electronics are projected to have very high value in small quantity. Moreover, in terms of gaining widespread scientific and commercial acceptance for the BlackLight Process, it is relatively easy to validate the properties of a chemical compound. A solid chemical compound is a product that can be examined directly and its existence proven unequivocally—it either exists or it doesn't. This also means that its patents are well defined and easy to defend. The products are much more diverse, so broad industry adoption is anticipated.

In addition to direct cell power to electric power conversion, thermal power from the plasma produced by the BlackLight Process may be converted to electricity by powering a turbine. Contemporary central station thermal generation systems have been optimized to match their respective thermal sources. Since BlackLight-technology is not combustion or nuclear, an opportunity exists to dramatically reduce the complexity of the generation station. The BlackLight Process may be used as a thermal source for central or distributed power through use of a modified steam or gas turbine. The BlackLight adaptation of the steam-based system replaces the heat source of the boiler with the gas cell. The BlackLight adaptation of the gas turbine replaces the combustor of a conventional machine with a gas cell and a heat exchanger incorporating the BlackLight Process. High pressure air from the compressor is heated by the BlackLight energy cell heat exchanger before expanding through the power turbine. The exhaust would contain no combustion products. With energy production from hydrogen at a hundred times combustion energy, fuel cost would become an inconsequential consideration, and refueling intervals would be consistent with other maintenance. Alternatively, an on-site electrolysis system producing hydrogen from water could provide unlimited fuel with periodic additions of small quantities of water.

A typical chemical plant is projected to produce 100 MW electric power as a side product. Power and chemical cells may be fabricated using readily available materials, and systems such as steam or gas turbine systems are scalable over a large range [e.g. distributed units (1 MW) to central power plants (1 GW)]. The projected cost for a combined chemical and energy plant is about $250/kW. The two functions could work seamlessly together and generate a dual income stream with a reduction of business risk. Rather than producing nuclear or fossil fuel waste which requires disposal, the BlackLight chemical plant will produce HHCs which have potential for far-reaching applications in many industries such as batteries for electric vehicles at significant earnings. For example, a 100 MW chemical plant is projected to produce $300 M in electric vehicle battery revenue from 200,000 batteries with $23 M from electricity sales at 3 ¢ kWh.

4. SOLUTION TO THE ENERGY PROBLEM?

The world's current energy system is unsustainable. Furthermore, the world's current energy system is not sufficiently reliable or affordable to support widespread economic growth. The productivity of one-third of the world's peoples is compromised by lack of access to commercial energy, and perhaps another third suffer economic hardships and insecurity due to unreliable energy supplies.[23]

Solar and wind power are prohibitively expensive. Billions of dollars have been spent to harness the energy of hydrogen through hot fusion using extremely hot plasmas created with enormous energy input using complex, expensive systems. By contrast, the Company's reactions indicate that over 100 times the energy of its combustion is released from hydrogen with the formation of a plasma as a by-product at relatively low temperatures with simple, inexpensive systems. And, in the Company's reactors, the plasma may be converted directly to electricity with high efficiency avoiding a heat engine such as a turbine. In addition, rather than producing radioactive waste, the BlackLight Process produces compounds having extraordinary properties. The implications are that a vast new energy source and a new field of hydrogen chemistry have been discovered.

The advantages of the BlackLight process over existing energy forms, such as fossil fuels and nuclear power, include: (1) the water, which is the fuel for the process, is safe and inexpensive to contain; (2) the reaction is prospectively easily controlled; and (3) the byproduct, HHCs, have great potential commercial value. The projection of the capital cost per kilowatt capacity of a gyrotron system may be an order of magnitude less than that of the typical capital cost for a fossil fuel system and two orders of magnitude less than that of the typical capital cost for a nuclear system. The power cell may also be interfaced with conventional steam-cycle or gas turbine equipment used in fossil fuel power plants. In either case, fuel costs are eliminated since the fuel, hydrogen, can be generated by a fraction of the electrical output power. The cost factors per kilowatt/hour are the capital, maintenance and operation costs of the gas cell and plant. These costs are further reduced by elimination of the costs of handling fossil fuels and managing the pollution of the air, water, and ground caused by the ash generated by fossil fuels.

4.1. BlackLight Distributed Generation

Central station generation and distribution, the mainstay of electrical power production for the last 100 years worldwide, is now being supplemented in an increasing number of areas by smaller power units closer to the end-user group. Most distributed-generation units are in the capacity range of 100 kW–3 MW (electric), but some could be as large as 250 MW (electric). Distributed generation solves some of centralized power's inherent problems of transmission and distribution line losses, electromagnetic pollution fears from high-tension lines, cost and difficulty of transmission-line maintenance, and inefficiencies in load factor design of power plants (wherein the use of a 20% capacity safety factor is still a common industry practice when estimating peak loading). The Company's technology may be ideal for distributed generation with significant reductions in grid complexity and generation capital equipment requirements.

The Company projects that the residential market may be broadly served by a 25 kW unit, and the commercial market may be broadly served by modular 1 MW units. This approach may replace the grid since in addition to avoidance of line losses, a major economic advantage of distributed power is the avoidance of transmission tariffs which could amount to 50% of the cost of electricity to a customer. Using BlackLight's distributed power generation technology, considerable savings can be realized by eliminating the transmission and distribution capital equipment, operations, and maintenance costs. Also, energy can be saved, given that electricity "demand" also includes substantial transmission and distribution losses from the traditional central-station type power generation systems. These considerations are important considerations for developing nations.

As the world's population grows from about 6 billion (in 1999) to an estimated 7 billion by 2010, most of the new energy demand will come from less-developed countries (LDCs), as these countries' living standards increase. LDC energy demand has long been answered by economic development programs generally aimed at the development of large, central-station power plants. These do not adequately address the thermal and lighting needs of the half the world's population which is poor, many of whom still use carbon fuels for these purposes. The solution for LDC's may be distributed power facilitated by BlackLight Power technology since no fuel, power plant, or transmission grid infrastructure is required.

5. BLACKLIGHT POWER TECHNOLOGY—A NEW PARADIGM IN ENERGY AND ELECTRICITY GENERATION

The products of the BlackLight Process are power, plasma, light, and novel HHCs. Using advanced catalysts in its gas power cell, the Company has sustained an energetic plasma in hydrogen at 1% of the theoretical or prior-known voltage requirement and with 1000's of times less power input in a system wherein the plasma reaction is controlled with a weak electric field. A plasma is a very hot, glowing, ionized gas. The plasma is produced from reactions which release energies over 100 times the energy of the combustion of hydrogen and correspond to an equivalent electron temperature of over 1,000,000 °C. The plasma produced in the Company's cells cannot be produced by any chemical reaction other than the Company's process.

Typically, a heat engine such as a turbine is used for converting heat into electricity. However, plasma power may be directly converted into electrical power. The technology is not based on heat. Thus, heat sinks such as a river or cooling towers as well as thermal pollution are largely eliminated. Based on research and development in this area of converters, the Company expects that routine engineering will result in devices that have

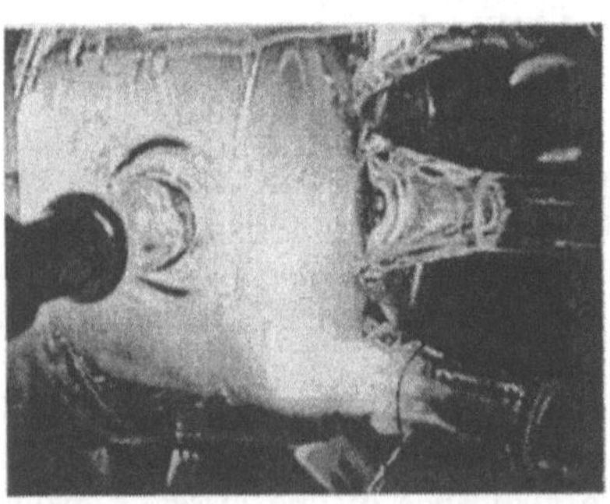

Figure 1. Plasma Generated by the BlackLight Process.

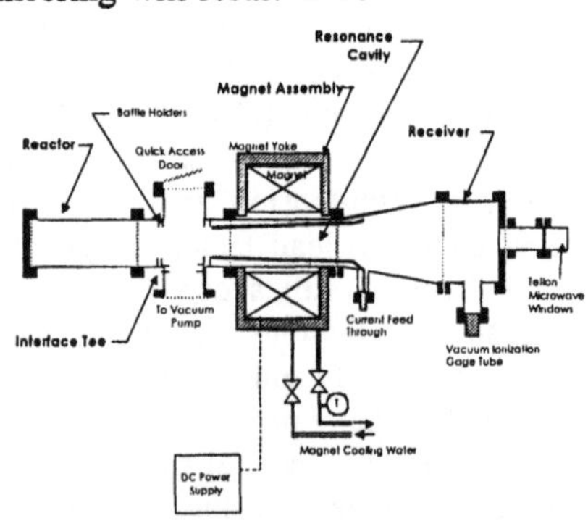

Figure 2. Gyrotron Schematic.

higher conversion efficiencies than turbines. The device is linearly scaleable from the size of hand held units to large units which could replace large turbines. And, unlike turbine technology wherein the cost per unit capacity soars with miniaturization, the Company anticipates that the unit cost per capacity will be insensitive to scale. The Company anticipates applications for its technology in broad markets such as premium power, microdistributed power, motive power, consumer electronics, portable electronics, telecommunications, aerospace, and uninterruptable, remote, and satellite power supplies.

Plasma may be directly converted into electricity using a device called a gyrotron which is established technology for converting energetic electrons into microwaves. Conventionally the source of energetic electrons comprises an electron beam or a plasma formed by electrical input such as a high voltage discharge. Prior to the development of the Company's technology, it was not possible to generate a plasma in hydrogen chemically. The BlackLight Process generates an energetic plasma in hydrogen which is *a new source of energy*.

The energy released by the catalysis of hydrogen to form HHCs produces a plasma in the cell. The energetic electrons of the plasma produced by the BlackLight Process are introduced into an axial magnetic field where they undergo cyclotron motion. The force on a charged ion in a magnetic field is perpendicular to both its velocity and the direction of the applied magnetic field. The electrons of the plasma orbit in a circular path in a plane transverse to the applied magnetic field for sufficient field strength at an ion cyclotron frequency ω_c that is independent of the electron velocity. Thus, a typical case which involves a large number of electrons with a distribution of velocities will be characterized by a unique cyclotron frequency that is only dependent on the electron charge to mass ratio and the strength of the applied magnetic field. There is no dependence on their velocities. The velocity distribution will, however, be reflected by a distribution of orbital radii. The electrons emit electromagnetic radiation with a maximum intensity at the cyclotron frequency. The velocity and radius of each electron may decrease due to loss of energy and a decrease of the temperature.

The gyrotron comprises a resonator cavity which has a dominate resonator mode at the cyclotron frequency. The plasma contains electrons with a range of energies and trajectories (momenta) and randomly distributed phases initially. Electromagnetic oscillations are generated from the electrons to produced induced radiation due to the grouping of electrons under the action of the self-consistent field produced by the electrons themselves with coherent radiation of the resulting packets. In this case, the device is a feedback oscillator. The theory of induced radiation of excited classical oscillators such as electrons under the action of an external field and its use in high-frequency electronics is described by A. Gaponov et al.[24] The electromagnetic radiation emitted from the electrons excites the mode of the cavity and is received by a resonant receiving antenna.

The radiated power and the power produced by the BlackLight Process may be matched such that a steady state of power production and power flow from the cell may be achieved. The rate of the hydrogen catalysis reaction may be controlled by controlling the total pressure, the atomic hydrogen pressure, the catalyst pressure, the particular catalyst, and the cell temperature. Very fast response times may be achieved by controlling the rate of reaction and plasma formation with an applied electric or magnetic field which influences the catalysis rate. Plasma and a gyrotron can respond essentially instantaneously. Thus, unprecedented load following capability is possible.

The gyrotron relies on established microwave technology which may achieve very high efficiencies (e.g. 80%) conversion of energetic electrons into microwaves.[25] A 0.1

Figure 2. Gyrotron Prototype.

Table 3. Economics of International Fuel Cells Corp.

Basis: Installed Cost < $1,000/kW DOE Credit—$3,000.kW
Capital Recovery Factor—12%
Annual Load Factor—95% (8,322 hrs of operation)
Electric Efficiency (higher heating value)—36%
Heat Rate—9,480 Btu / kWh
Waste Heat Recovery as Hot Water
(Equivalent to 875,000 Btu/hr of fuel input at 80% efficiency)
Implicit Overall Thermal Efficiency—82%
Natural Gas Cost—$3.50 / million Btu

Cents/kWh	
Capital Charges	4.3
Fuel	3.3
O&M	2.0*
Subtotal	9.6
Hot Water Credit	−1.5
Net Power Cost	8.1

* Includes $600/kW overhaul costs every six years

Tesla magnetic field will produce about 2.5 GHz microwaves. The microwaves are then rectified into DC electricity. Rectification efficiency at 2.5 GHz is about 95%.[26-29] The DC electricity may be inverted and transformed into any desired voltage and frequency with conventional power conditioning equipment.

The plasma formed by the BlackLight Process and the gyrotron have been tested independently. Current work is in progress on testing gyrotron powered by the BlackLight Process generated plasma.

5.1. Power Balance Analysis

The commercial unit would comprises a 3-stage power generator. Stage 1 would be electrolysis to provide hydrogen fuel; stage 2—production of plasma in a gas cell; and stage 3—conversion of plasma to microwaves to electricity.

Using even relatively conservative assumptions for reaction yield and power density, a competitive power generation unit appears easily possible: (1) Production of about 100 times electrical power as electrolysis power; (2) Production of *green* emission (oxygen only) *zero* CO_2 *emission;* (3) No fossil-fuel combustion by-products; (4) Essentially no waste heat since the gyrotron is not a heat engine; (5) Tremendously more efficient at energy conversion to electricity; and (5) Projected to dominate the home and microdistributed markets.

5.2. Comparison with Competing Microdistributed Technologies

The Company's process may start with water as the hydrogen source and convert it to HHCs; whereas, fuel cells typically require a hydrocarbon fuel and an expensive reformer to convert hydrocarbons to hydrogen and carbon dioxide. The Company's plasma to electric conversion technology with no reformer, no fuel cost, creation of a valuable chemical by-product, and significantly lower capital costs and O&M costs are anticipated to result in household units that are competitive with central power and significantly superior to competing microdistributed power technology such as fuel cells and micro combustion turbines. With a focus on large scale production of microdistributed devices, the Company anticipates rapid penetration of the electricity energy market. In this case, the Company plans to form strategic alliances with component manufacturers, systems assemblers, and service companies to provide power for consumers with units under lease or by sale. The service companies may be utilities. Other services or utility companies such as water, gas, telephone, cable, plumbing, and HVAC companies are also potential partners. The Company may have its plasma-electric power cell manufactured under contract or license. Alternatively, the Company may manufacture the units itself.

Some of the competitive advantages of BlackLight Power generation over the competing microdistributed technologies fuel cells and micro combustion turbines are no fuel costs, no fuel handling issues nor pollution, not a heat engine and not electrochemical, no reformer, solid state device, chemically-generated plasma with proven microwave technology, linearly scalable, cost competitive (lower capital and O&M costs), long product lifetime, appliance-like, load following, no grid connection (gas or electric for fuel or load leveling), high 9's power capability, closed system, and valuable solid chemical by-product.

With strategic alliances, the Company plans to develop, manufacture, and market a unit of approximately 25 kWe which is a desirable size for a modular uninterruptable power supply for the premium power market. 25 kWe is also capable of providing for the total power requirements of a single family residence or a light commercial load. The potential advantages of the Company's power system compared to fuel cells are (1) zero fuel costs, (2) capital and O&M costs that are 10% that of fuel cells, and (3) valuable chemicals are produced rather than pollutants such as carbon dioxide. Thereby, the cost per kilowatt of electric generated by the Company's plasma-electric power cell is projected to be about 10% of that of a fuel cell. In addition, an energy consumer may also derive revenue by selling power back onto the distribution system when the full capacity of the system is not required by such consumer.

The only mass manufacturable components required to produce a gyrotron system are a magnet, a resonator cavity or waveguide, and a antenna-rectifier. For implementation in the third world and acquisition of market share in the first world, the plasma-electric cell requires essentially no fuel and fuel distribution infrastructure, no regional or on-site pipelines, no utility connection (gas or electrical), no electric lines, and no specialized or centralized manufacturing expertise. In each category, competing technologies are at a competitive disadvantage which could prevent broad adoption even if they were viable based on logistics and costs.

Fuel cells are not cost competitive with BlackLight technology. The cost of electricity with a molten carbonate fuel cell which has a much lower capital cost compared to a proton exchange membrane fuel (PEM). The projected capital cost for a BlackLight 5–25 kW plasma-electric system is as follows: rectifier—$200; inverter—$200, permanent magnet—$150, and cell—$100, totaling $650.

Also with strategic alliances, the Company further plans to develop, manufacture, and market a unit of approximately 1 MWe. One to ten of these units should provide the total power load requirements of a central power grid substation. The potential advantages of the Company's power system compared to central power are the same as with plasma-electric power cell. The cost per kilowatt of electric generated by the Company's plasma-electric power cell is projected to be about 20% of that of central power (see Table 2). With the installation of substation units, light commercial, and residential units, all components of the present central power generation infrastructure upstream from the substation may be eliminated. Some infrastructure components that may be eliminated by the Company's technology with associated cost savings are: (1) high voltage transformers, (2) high voltage transmission lines, and (3) central power plants, including their associated turbines, fuel and pollution handling systems, ash, pollution, coal trains, coal mines, gas pipelines, gas fields, super tankers, oil fields, nuclear power plants, uranium processing plants, and uranium mines.

5.3. Motive Power—Plasma-Electric and Battery

The capital cost for BlackLight power for motive power are comparable to the cost of an automotive internal combustion engine. Whereas, fuel cells are two orders of magnitude too expensive and require trillions of dollars to be invested in a hydrocarbon to hydrogen refueling system. In contrast, a motive power plant based on BlackLight technology uses water as the fuel and requires no infrastructure. The Company is considering several promising options to commercialize its process in the motive power market. In addition to stationary power, the plasma-electric system may be used for motive power. The Company is also developing a high voltage battery which may power an electric vehicle.

5.4. Conclusion

The BlackLight Process has potentially very broad applications including: electrical power generation, space and process heating, motive power, and production of HHCs.

The technology generates plasma and heat from hydrogen, which may be obtained from ordinary water. The implications of this development could be significant. If the technology becomes proven, then the energy from this process could possibly be used to cleanly and cheaply meet the world's demand for thermal, chemical, and mechanical energy as well as electricity. Over time, it may be possible to replace or retrofit coal-fired, gas-fired, and oil-fired electric power plants. This would help to abate global warming and air and water pollution. Moreover, it may be possible to replace or retrofit some of the world's nuclear power plants. With BlackLight technology, an opportunity exists to dramatically reduce the complexity and the cost of the generation station, which includes fuel handling, thermal generation, thermal to electrical conversion, pollution abatement and spent fuel disposal or storage systems.

The Company is focusing on possible electrical and heating applications for its technology including a fit with a converter to make electricity. Electrical power generation with the Company's plasma-electric power technology may represent a major opportunity to use a microdistributed system to replace existing infrastructure at considerable savings in capital and generation costs. Residential/light commercial units, substation units, and a low voltage local distribution system could replace the central power based current system. Adaptation of the Company's technology is facilitated by the deregulation of the utility industry.

6. REFERENCES

1. R. Mills, *The Grand Unified Theory of Classical Quantum Mechanics*, January 2000 Edition, BlackLight Power, Inc., Cranbury, New Jersey, Distributed by Amazon.com.

2. R. Mills, M. Nansteel, and Y. Lu, Anomalous hydrogen/strontium discharge, *European Journal of Physics D*, submitted.
3. R. Mills, J. Dong, Y. Lu, Observation of extreme ultraviolet hydrogen emission from incandescently heated hydrogen gas with certain catalysts, *Int. J. Hydrogen Energy*, **25**, 919-943 (2000).
4. R. Mills, Observation of extreme ultraviolet emission from hydrogen-KI plasmas produced by a hollow cathode discharge, *Int. J. Hydrogen Energy*, in press.
5. R. Mills, Temporal behavior of light-emission in the visible spectral range from a Ti-K_2CO_3-H-cell, *Int. J. Hydrogen Energy*, in press.
6. R. Mills, Y. Lu, and T. Onuma, Formation of a hdrogen pasma from an incandescently heated hydrogen-catalyst gas mixture with an anomalous afterglow duration, *Int. J. Hydrogen Energy*, in press.
7. R. Mills, M. Nansteel, and Y. Lu, Observation of extreme ultraviolet hydrogen emission from incandescently heated hydrogen gas with strontium that produced an anomalous optically measured power balance, *Int. J. Hydrogen Energy*, in press.
8. R. Mills, B. Dhandapani, N. Greenig, J. He, Synthesis and characterization of potassium iodo hydride, *Int. J. of Hydrogen Energy*, **25**, 1185-1203 (2000).
9. R. Mills, Novel Inorganic Hydride, *Int. J. of Hydrogen Energy*, **25**, 669-683 (2000).
10. R. Mills, Novel hydrogen compounds from a potassium carbonate electrolytic cell, *Fusion Technology*, **37**(2) 157-182 (2000).
11. R. Mills, B. Dhandapani, M. Nansteel, J. He, T. Shannon, A. Echezuria, Synthesis and characterization of novel hydride compounds, *Int. J. of Hydrogen Energy*, in press.
12. R. Mills, Highly stable novel inorganic hydrides, *Journal of Materials Research*, submitted.
13. R. Mills, B. Dhandapani, M. Nansteel, J. He, A. Voigt, Identification of compounds containing novel hydride ions by nuclear magnetic resonance spectroscopy, *Int. J. Hydrogen Energy*, submitted.
14. R. Mills, The nature of free electrons in superfluid helium—a test of quantum mechanics and a basis to review its foundations and make a comparison to classical theory, *Int. J. Hydrogen Energy*, submitted.
15. R. Mills, N. Greenig, S. Hicks, Y. Lu, T. Onuma, Spectroscopic identification of a novel catalytic reaction of atomic hydrogen and the hydride ion product, in progress.
16. David R. Linde, *CRC Handbook of Chemistry and Physics*, 79 th Edition, CRC Press, Boca Raton, Florida, (1998-9), p. 10-175 to p. 10-177.
17. J. Phillips, J. Smith, S. Kurtz, Report On Calorimetric Investigations Of Gas-Phase Catalyzed Hydrino Formation, Final Report for Period October-December 1996, January 1, 1997, A Confidential Report submitted to BlackLight Power, Inc. provided by BlackLight Power, Inc., 493 Old Trenton Road, Cranbury, NJ, 08512.
18. USDOE 1999 Report
19. Merrill Lynch Report on Plug Power and Fuel Cell Market Size. December 6, 1999
20. S. Sanders, J. Chumbler, M. P. Zhang, Powering the Digital Economy/Digital Power Demand Meets Industrial Power Supply/Emerging Power Technologies for the Next 100 Years, Published by Stephens Inc. Investment Bankers, 111 Center Street, Little Rock, Arkansas, 72201, August, (2000).
21. T. Cooper, H. Harvey, Power Electronics, "Power Semiconductors and Power Supplies-The Building Blocks of the Digital Power Revolution, Published by Stephens Inc. Investment Bankers, 111 Center Street, Little Rock, Arkansas, 72201, (2000).
22. P. Huber, M. Mills, The Huber Mills Digital Power Report, *Powering the Telecosm*, Gilder Publishing, June 2000, Issue 3.
23. World Energy Assessment, http://services.sciencewise.com/content/index.cfm?objectid=309.
24. A. Gaponov, M. I. Petelin, V. K. Yulpatov, Izvestiya VUZ. Radiofizika, 10(9-10) 1414-1453 (1965).
25. V. A. Flyagin, A. V. Gaponov, M. I. Petelin, and V. K. Yulpatov, IEEE Transactions on Microwave Theory and Techniques, Vol. MTT-25, No. 6, June (1977), pp. 514-521.
26. R. M. Dickinson, Performance of a high-power, 2.388 GHz receiving array in wireless power transmission over 1.5 km, in 1976 IEEE MTT-S International Microwave Symposium, (1976), pp. 139-141.
27. R. M. Dickinson, *Bill Brown's Distinguished Career*, http://www.mtt.org/awards/WCB's%20distinquished %20 career.htm.
28. J. O. McSpadden, Wireless power transmission demonstration, Texas A&M University, http://www.tsgc.utexas.edu/power/general/wpt.html; *History of microwave power transmission before 1980*, http://rasc5.kurasc.kyoto-u.ac.jp/docs/plasma-group/sps/history2-e.html.
29. J. O. McSpadden, R. M. Dickson, L. Fan, K. Chang, A novel oscillating rectenna for wireless microwave power transmission, Texas A&M University, Jet Propulsion Laboratory, Pasadena, CA, http://www.tamu.edu, Microwave Engineering Department.

GAS RESOURCES FOR THE 21ST CENTURY

JACQUES MAIRE
CHAIRMAN, INSTITUT FRANÇAIS DE L'ENERGIE
HONORARY PRESIDENT, GAZ DE FRANCE

Mr Chairman,
Ladies and Gentlemen,
Friends and Colleagues,

I was greatly honoured to receive an invitation to speak to you today about the probable role of natural gas over the coming decades. Thanks to my long experience at Gaz de France, the conference organizers thought I might have some interesting insight to share with you on this question. May I thank them most sincerely for giving me this opportunity both to present my viewpoint and to visit your beautiful state of Florida.

Over the last decade, climate change has become an issue of major concern for our fellow citizens and our governments. The problem of global warming due to the increasing concentration of certain gases in our atmosphere is no longer a purely scientific question. With the United Nations Framework Agreement on Climate Change in 1992 and the Kyoto Protocol in 1997, it has now entered the international political arena. On the basis of the precautionary principle, a new notion in the field of international negotiation, our governments have decided that the industrialized nations should reduce their emissions of greenhouse gases, carbon dioxide in particular.

Of course, as this conference is seeking to point out, many questions remain unanswered. The reality of global warming is no longer in doubt, though there is some discussion about its intensity. But the scientific community does not agree about its causes and its potential effects are another subject of debate.

Clearly, the signature of the Kyoto protocol in 1997 was above all a political act. Our political leaders have realized that, faced with a little understood but potentially threatening risk, we must not bury our head in the sand. The most industrialized countries (the so-called Annex B countries), held responsible for the increasing atmospheric concentration of carbon dioxide and other greenhouse gases over the last two centuries, have pledged to lower their emissions by 5.2% on average, with respect to 1990 levels, over the period 2008-2012.

This calls for two remarks:

Global Warming and Energy Policy, Edited by Kursunoglu *et al.*
Kluwer Academic/Plenum Publishers, New York 2001

- Of all the international environmental treaties, it is certainly the Kyoto protocol that will have the greatest impact on our economies. Indeed, it concerns a key component of our economic development, namely energy. The main greenhouse gas is carbon dioxide. More than 80% of anthropogenic carbon dioxide emissions (i.e., due to human activity) are linked to the use of energy, through the combustion of coal and other solid fuels, oil and its derivatives and natural gas. Measures to reduce greenhouse gas emissions will affect the way we use energy and hence the very conditions of our development. How can we respond effectively and fairly to this problem without threatening our way of life and without compromising the development aspirations of four-fifths of humanity? In this respect, the flexibility mechanisms, whose principles were defined in the protocol, are fundamental [*NDR : voir, juste avant l'intervention, si l'on peut ajouter quelques mots sur les résultats de COP6*]
- My second remark concerns the timing of our response. Officially, the Kyoto protocol is not yet in force, as the prescribed conditions (i.e., the number of countries which have ratified it and their characteristics) have not been satisfied. However, it would be short-sighted, in my view, to use the hypothetical failure of the protocol as an excuse for inaction. In the field of energy, change takes time. The weight of existing energy infrastructures and investments creates strong inertia. If we want to bring down greenhouse emissions a decade from now, we must take action straight away, giving priority, of course, to all actions whose consequences, even if they do not achieve the desired results, have minimum negative impact. The reduction targets set in Kyoto appear modest in absolute terms, but are in fact highly ambitious. If we do not act quickly they will certainly be impossible to achieve.

In this context, natural gas has a key role to play. All experts agree that natural gas will be the number one energy choice in the first decades of the 21st century. There are two main reasons for this.

1. Natural gas offers key environmental advantages. It is the fossil fuel which emits the least carbon dioxide during combustion, emitting 25 to 30% less than petroleum products and 40 to 50% less than solid fuels for the same quantity of energy produced. Moreover, natural gas combustion emits no sulfur compounds or solid particles and very few nitrogen oxides.
2. Natural gas utilization technologies have made great progress over the last ten years, to the benefit of all sectors using natural gas as an energy source. Let me give you three examples. The efficiency of gas boilers available on the domestic and commercial markets fifteen years ago reached 80% at most; today, condensing boilers are able to recover a part of the latent heat of water vaporization, with an efficiency of more than 100% (net calorific value). A second example: combined-cycle gas turbines are the most economical and least polluting technology for large-scale power generation. Efficiencies of more than 55% are now possible, coupled with a substantial reduction in pollutant emissions per unit of energy produced. My third example, in a field once reserved for liquid fuels, concerns the use of gas as an automotive fuel. Natural gas for vehicles is developing rapidly in many countries and helping to clean up the air in our towns and cities.

These spectacular technological advances will continue in the 21st century, thanks to the sustained research and development efforts of the gas industry to ensure that the qualities of natural gas are exploited to the full.

But some people point out that methane itself is a greenhouse gas, and they are right, since one tonne of methane has an effect equivalent to 21 tonnes of CO_2. But the production and utilization of natural gas account for only a small share (7%) of the methane emitted into the atmosphere, and careful grid management keeps emissions to a minimum. There are major geographical variations in gas industry methane emissions, with one third produced in North America and 20% in the former USSR. Outside the gas industry, the other methane emissions are attributable to the fact that methane is a fossil energy, but also a renewable energy (!) produced by landfills (33% of emissions), from which it is partly recoverable, by cattle and by rice paddies.

This "dash for gas" will mean a sharp rise in demand. Among the many energy development scenarios put forward by expert bodies, those presented by the International Gas Union (IGU) at its recent World Gas Conference in Nice in June 2000 provide, in my view, the best synthesis of current data. The IGU predicts that world natural gas demand will more than double by 2030, rising to 4,700 billion cubic meters from 2,300 billion cubic meters today. This corresponds to a mean annual growth rate of 2.3%, compared to a forecast growth rate of 1.7% for total energy demand over the same period. Of the 2,400 billion cubic meters of additional gas demand, more than 1,000 billion cubic meters (42%) could be used for power generation, the fastest growing market over the next thirty years.

The geographical distribution of gas demand will also change. Today, the cumulative consumptions of North America, Europe and the former USSR represent three-quarters of world demand. By 2030, these three regions will account for only 62% of the gas consumed in the world and the rate of growth in gas demand will be 2 to 2.5 times lower than in the other regions, which include all emerging and developing countries.

So the gas industry is optimistic about future demand for its product. But will it be able to satisfy this demand? Are natural gas reserves sufficient to justify this optimism?

The gas industry's answer is clear and unambiguous: world gas reserves are more than sufficient to meet the predicted growth in demand. Here again, I am basing my affirmations on the IGU study presented in Nice. This is not the first prospective study by the IGU in this field, so it is possible to make comparisons. These comparisons reveal a steady increase over time, both in proven reserves and in total reserves.

Proven reserves are the volumes of discovered natural gas that could be produced under current economic conditions and with available technologies. Between the 1988 estimates and those presented at the recent IGU Conference, these proven reserves have increased by almost 50%, rising from 104,000 to 153,000 billion cubic meters, despite a cumulative production of 25,000 billion cubic meters over this period. This increase is due not only to sustained exploration activities in all regions of the world, but also to technological developments that have made it possible to re-evaluate existing reserves. Proven reserves are sufficient to satisfy demand for sixty-six years at current production levels. The figure was about fifty years in the previous studies.

These proven reserves certainly do not represent the ultimate quantity of natural gas that can be recovered on our planet. Firstly, as I already mentioned, the amount of available gas is increasing steadily thanks to technological progress. Secondly, many

Fossil fuels and the Environment

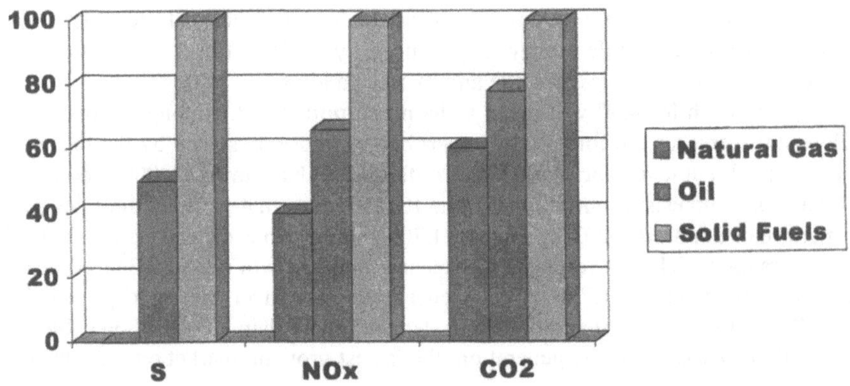

Gas Demand: IGU Base Case Scenario

bcm	1998	2010	2030	2030/1998 %/ year
Africa	54	100	170	3.6
Middle East	183	300	510	3.3
Asia & Oceania	263	470	770	3.4
East Europe & North Asia	530	660	990	2.0
West & Central Europe	423	580	690	1.5
North America	721	950	1,250	1.7
South America	82	160	320	4.3
World	**2,256**	**3,220**	**4,700**	**2.3**

Gas Demand: IGU Alternative Scenario

bcm	1998	2010	2030	2030/1998 %/year
Africa	54	90	140	3.0
Middle East	183	250	380	3.2
Asia & Oceania	263	430	640	2.8
East Europe & North Asia	530	640	820	1.4
West & Central Europe	423	530	600	1.1
North America	721	790	1,040	1.2
South America	82	120	180	2.5
World	**2,256**	**2,850**	**3,800**	**1.6**

Source: IGU World Gas Conference 2000

Proven Reserves by region

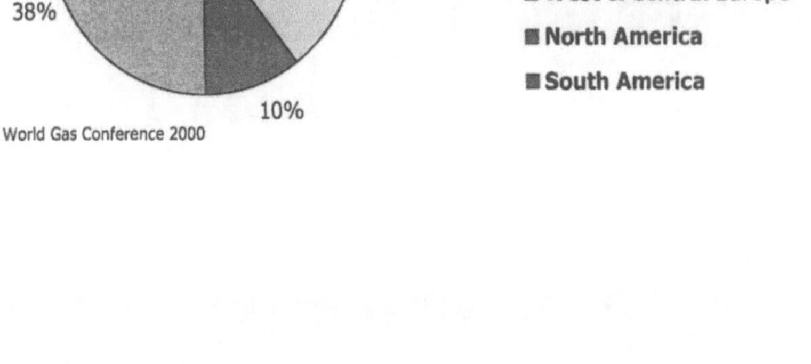

Source: IGU World Gas Conference 2000

Proven and Total Reserves by region (bcm)

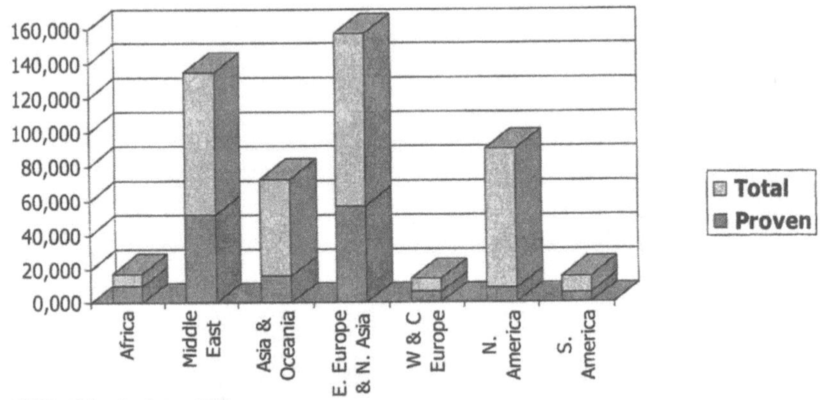

Source: IGU World Gas Conference 2000

Ratio reserves / production by region (years)

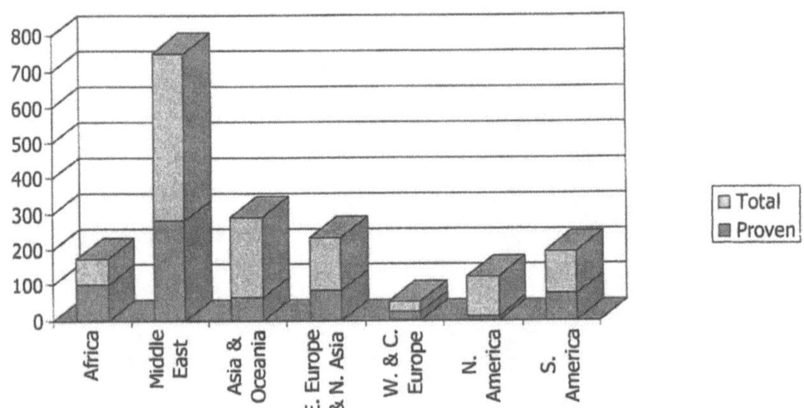

Source: IGU World Gas Conference 2000

Interregional trade

bcm	1997	2030
LNG	108	250 / 370
Pipes	134	380 / 460
Total	242	630 / 830

Source: IGU World Gas Conference 2000

Gas Infrastructures

	1997	2030
Transmission Pipelines (1000 km)	1078	1665 / 2045
Liquefaction Plants (MT/year)	91	165 / 375
Methane Tankers (bcm)	10	24 / 44
Gasification Plants (bcm/year)	264	305 / 320
Underground Storages (bcm)	302	545 / 695
Distribution Networks (1000 km)	4387	6980 / 8870

Source: IGU World Gas Conference 2000

Investments (renewal & development) from 1998 to 2030

Billion US$

Transmission Pipelines	680 / 838
Liquefaction Plants	60 / 137
Methane Tankers	40 / 73
Gasification Plants	25 / 42
Underground Storages	140 / 190
Distribution Networks	1059 / 1337
Total	2004 / 2616

Source: IGU World Gas Conference 2000

geographical zones have not yet been explored. Lastly, methane can also be recovered from unconventional sources, some of which are already being exploited. One example is coal-bed methane. Total reserves are estimated at around 229,000 billion cubic meters, mainly in North America, the former USSR and in China. Methane can now also be recovered from low permeability fields and their industrial development has already begun. These reserves could total 175,000 billion cubic meters, two-thirds of which are located in the Middle East and the former USSR. Other unconventional sources are known but have not yet been exploited on an industrial scale. This is the case for natural gas hydrates. Large areas of our ocean floors are covered by a layer of frozen gas with a very high methane content. We do not yet have the technologies to exploit this resource, though its existence is certain and estimates of its total volume are colossal – more than 8 million billion cubic meters! Lastly, certain experts believe that immense gas reserves exist deep down in the earth under our feet. Unlike the hydrocarbons we use today, this hypothetical gas is not organic, but of igneous origin. However, there is still no proof that igneous methane really exists.

On the basis of conservative estimates, including only a small share of these unconventional resources, the IGU has calculated that total reserves represent a volume of 500,000 billion cubic meters, enough for 220 years of production at current levels. No wonder the gas industry is optimistic about its future! Last but not least, waste due to the use of inefficient technologies or poor commercial management is another potential resource, difficult to classify as either conventional or unconventional. People sometimes joke that the largest Russian gas reserve is not in Siberia but in Moscow – and they are not far from the truth! This potential is difficult to evaluate and perhaps even more difficult, on a political level, to mobilize, but it certainly does exist.

So there is a good match between potential supply and potential demand at world level. For the next fifty years at least, natural gas supply should be more than sufficient to satisfy demand even if, as predicted, consumption increases very quickly. But, as I said, this is a global assessment. Unfortunately, this overall balance is not so perfect when we examine the situation at the regional level. The world's natural gas reserves are very unevenly distributed and their geographical location does not coincide with the centers of future demand.

I said earlier that by 2030, 62% of gas consumption would be concentrated in North America (26.5%), in the former USSR (21.1%) and in Europe (14.6%). In 2030, consumption in the Asia/Oceania region will be higher than in Europe.

Most of the world's proven natural gas reserves are located in two regions: 37% in Eastern Europe and Northern Asia and 33% in the Middle East. North America possesses only 4% of reserves. For total reserves, the situation is slightly different: if we include North America's immense unconventional reserves, its share of total world reserves increases to 18%, but remains well below the share of the former USSR (31%) or the Middle East (27%).

These regional disparities can also be illustrated by looking at the ratio of proven reserves to current production. In the Middle East, this ratio exceeds 250 years, compared to less than 20 years in North America and slightly more in Western Europe. The same disparity is observed in the ratio of total reserves to current production, which stands at 700 years in the Middle East, 120 years in North America and 50 years in Western Europe.

This comparison of the geographical distribution of future demand and of gas reserves shows that natural gas sources will become increasingly distant from the points of consumption. This will mean an increase in natural gas trade between the different regions of the world. Today, this trade totals 242 billion cubic meters, 45% in the form of LNG and 55% as pipeline gas. According to the IGU's prospective scenarios, the annual volumes traded between the major regions of the world will total 630 to 830 billion cubic meters by 2030, between 2.6 and 3.4 times the current volume, with the share of LNG remaining stable at around 45%.

To trade gas and supply consumer markets, the gas industry has developed extensive infrastructures: transmission pipelines, liquefaction plants, methane carriers, regasification plants, underground storage facilities and distribution networks. The value as new of these installations is estimated at 1,300 billion dollars.

With the foreseeable increase in natural gas demand and trade, many new gas infrastructures will be needed. Over the next thirty years, transmission, storage and distribution infrastructures will need to double in size and existing installations will need to be totally or partially replaced.

Hence, by 2030, gas transmission pipelines will reach a total length of between 1.7 and 2 million kilometers; liquefaction installations could produce between 165 and 375 million tonnes of LNG per year; the transport capacity of LNG carriers will increase to between 24 and 45 million cubic meters; underground storage reservoirs will have a working gas capacity of between 550 and 700 billion cubic meters and the length of distribution networks will reach between 7 and 9 million kilometers.

The IGU estimates that development investments of between 800 and 1,400 billion dollars will be needed over the period 1998-2030, of which almost half (400 to 650 billion dollars) will be devoted to distribution networks. The rest will concern pipeline transmission systems (300 to 400 billion dollars), LNG installations (between 70 and 200 billion dollars) and storage facilities (80 to 130 billion dollars). On top of these development investments, around 1,200 billion dollars will be needed to replace existing infrastructures.

Altogether, the gas industry will invest between 2,000 and 2,600 billion dollars between 1998 and 2030 for infrastructure development and replacement (excluding gas exploration-production investments). More than 70% of these investments will be concentrated in three regions: North America, Europe and the former USSR. Though the sums involved sound colossal, they represent only 10% of the gas industry's provisional sales and less than 0.2% of the world GDP over this period. It should not, therefore, be too difficult for our industry to mobilize the necessary capital.

But we should not be overoptimistic. The intrinsic qualities of gas and the efforts of the gas industry alone will not be enough to prevent the golden carriage from turning into a pumpkin. A favourable economic and regulatory environment is also essential. The legal systems in place must give our industry the freedom to cover its overall costs through its income from gas sales. The other systems, which unfortunately still exist in some places, often lead to operating difficulties over the long term, on the political and social levels especially. Moreover, given the scale, the duration and the fixed nature of gas investments, long-term economic and regulatory visibility is vital. This does not mean that the environment must be the same on all markets and under all circumstances. For the mature gas markets, fair competition, coupled with effective regulation mechanisms are the preconditions for harmonious development. This is the situation which exists or is now being set up in North America and the European Union. But for emerging gas markets, certain forms of monopoly are vital, for a time at least, to attract and protect the external investors whose contribution is often essential. There is no single model.

As I reach the end of my presentation, I hope to have convinced you that the gas industry's optimism about its future is not unfounded. Thanks to the environmental qualities of natural gas and its increasingly energy-efficient utilizations, and thanks to the competence and know-how of the gas industry, natural gas will be a key component of sustainable development in the 21st century. Let us hope that its contribution is not held back or compromised by a poor understanding, on the part of players outside our industry, of the conditions required for its development.

Mr Chairman, ladies and gentlemen, thank you for your attention.

The Future of Nuclear Energy in the U. S.

Angelina S. Howard*

Good afternoon, and thank you for the opportunity to participate in this discussion on the future of nuclear energy and the environment. I'm delighted to be in Fort Lauderdale—or anywhere in Florida—as winter begins its descent on the northeast. It isn't quite beach weather, but I'll take it!

By way of introduction, I thought I would tell you a little bit about my organization, the Nuclear Energy Institute.

NEI develops public policy for its more than 270 national and international members of the nuclear industry. In addition to representing every U.S. utility that operates a nuclear power plant, NEI's membership includes nuclear fuel cycle companies, suppliers, engineering and consulting firms, national research laboratories, universities, labor unions, law firms and manufacturers of radiopharmaceuticals. Our role is to focus the collective strength of our membership—about 18 percent of which are international companies—in order to help shape policies in the U.S. and globally that foster the use of a wide range of nuclear technologies.

In short, our role is to provide the policy direction and leadership to advance this remarkable industry and help provide a strategic direction to what we feel is a renaissance for these technologies.

This afternoon I will talk about nuclear energy mainly from a U.S. perception and its role in establishing U.S. energy diversity. I'll also address its historical and ongoing role in helping to mitigate the impact of air pollution in the United States—particularly in the areas around our most densely populated and economically vital urban areas. These areas need bulk supplies of electricity with limited environmental impact—because all generation types have some environmental impact.

I also will discuss the growing awareness that nuclear energy is critical to meeting future electrical demand. Nuclear energy is a vital component of any viable strategy to foster economic progress and improve the environment. In short, nuclear energy has an important role to play in the future.

My organization is truly optimistic about the future. We need only to take a look at the state of the nuclear energy industry in the United States to see why. Last year, for example, U.S. nuclear plants set a performance record by generating 728 billion kilowatt-hours of electricity. That is 54 billion kilowatt-hours more than the previous year and 152 billion kilowatt-hours more than in 1990. To put that in perspective, that is 20 percent of U.S. electrical demand, or approximately the equivalent of the combined nuclear generation of France, Japan and Belgium. It also is equal to all of the electricity demand of France and the United Kingdom combined.

---

*Angelina S. Howard, Executive Vice President, Nuclear Energy Institute, Washington, DC 20006-3708.

The pace of generation remains very high this year, as well. From January through July, U.S. nuclear power plants generated 442 billion kilowatt-hours of electricity, which is 6.8 percent above the same point last year.

Nuclear plants in the U.S also operated extremely efficiently last year. Capacity factors—the measure of the amount of electricity produced by each plant compared with the maximum possible—averaged 86.8 percent in 1999 for the 103 nuclear generating units. That is the highest average capacity factor in the world.

By contrast, in 1980 U.S. nuclear units had a net capacity factor of 57.6 percent and in 1990, 67.5 percent. The dramatic increase in the efficiency of America's nuclear stations should be viewed as one the most successful energy efficiency programs of the last decade. In fact, it was equivalent to adding nineteen 1,000-megawatt emission-free power plants to our nation's electricity grid. Remember this fact the next time you hear someone say that the United States is not building any new nuclear power plants.

U.S. plants are also operating more economically. Total costs are averaging 2.0-2.5 cents per kilowatt-hour. By comparison, the average costs for a new gas-fired combined cycle plant are 3.0-3.5 cents per kilowatt-hour. And that was before gas prices recently reached their all-time highs of well over $4 per million BTU.

Equally important, as the electric utility industry restructures to prepare for competition, U.S. industry leaders decided that it makes great business sense to re-license their nuclear units. Five units have received Nuclear Regulatory Commission approval for another 20 years of operation beyond their original 40-year license, for a total of 60 years. Five units have formal applications under NRC review, and 28 other units have informed the NRC of their intent to pursue license renewal.

That means, to date, the owners of 32 percent of the U.S. reactor fleet have decided that license renewal make competitive sense in a restructured market—and this is just the beginning.

Where do we go from here? Perhaps one of the most exciting areas for the industry is exploring the conditions for new nuclear plant construction.

A key driver of this effort is electricity demand growth—and a need for new generation—far beyond what anyone thought it would be a few short years ago. Actual demand growth in the first six months of 2000 was almost double what was projected last year. Overall energy prices and the doubling of natural gas prices have also have a major effect on the potential for new nuclear generation. Other factors driving the new plant discussions include the emergence of large nuclear generation companies and groups capable of making large capital investments—plus increasing clean air requirements.

In addition to its proven economic performance, nuclear energy's intrinsic emission-free nature continues to help mitigate air pollution and the potential for climate change.

In 1999, nuclear energy avoided emissions of 168 million metric tons of carbon, 4 million tons of sulfur dioxide and 2 million tons of nitrogen oxide. That means from 1973-1999, nuclear energy avoided 2.6 billion metric tons of carbon, 62 million tons of sulfur dioxide and 32 million tons of nitrogen oxide. In short, the increased number of plants and their enhanced performance avoided the environmental disruptions and impacts that would have occurred had companies brought equivalent conventional generation on line to meet energy needs.

By supplanting more traditional and price-sensitive fossil generation, nuclear energy added enormously to U.S. energy security and fuel diversity. At the time of the first OPEC oil embargo in 1973, approximately 20 percent of U.S. electricity supply came from oil-fired power plants and just five percent came from nuclear power. In the subsequent decades, 89 new nuclear units began operating—effectively replacing oil as a fuel source for electricity in the United States.

Today, nuclear power continues to provide a reliable hedge against volatile fuel prices and other energy supply disruptions, providing American businesses, homes, hospitals and schools with a reliable supply of electricity and protection from fluctuating fuel costs.

What is less well appreciated is that nuclear energy plants—from their first operation—were also seen as a vital hedge against a growing air pollution problem in the United States. In 1968, before a subcommittee of the U.S. House Committee on Science [and Astronautics], Dr. Joseph Lieberman, assistant director for nuclear safety at the Atomic Energy Commission, had this to say about nuclear energy:

> Another advantage of nuclear power plants is that there has been a growing awareness of their advantages as clean energy sources of power, which do not contribute to the current burden of air pollution...In fact, some utilities have chosen nuclear power and have indicated that in so doing, they wish to reduce air pollution.

Further confirmation of this notion comes from an interview with Phillip Fleger, the former chairman of Duquesne Light Company. He presided over the company's construction and operation of the first U.S. nuclear plant in 1955 near Pittsburgh, Pennsylvania. According to Fleger, his company wanted to build a coal-fired plant to help power Pittsburgh's urban redevelopment. The only problem was that the city's past as the center for iron and steel production in the U.S. had earned it the title of "Smokey City." Consequently, strict air quality controls—and strong public objection to more smoke—made the decision to pursue nuclear power an easy one. As Fleger said, the basic reason his company opted for nuclear power was pollution control.

In other words, from their inception, nuclear power plants were high-technology investments in the future integrity of the environment.

A similar situation exists today. In the U.S., the principal federal statute addressing air quality and man-made emissions is the Clean Air Act. The act sets pollution concentration levels allowable in the ambient air for sulfur dioxide, ozone, nitrogen oxide and particulate matter—not unlike the European Union's 1999 Gothenburg Protocol on air pollution reduction. Regulations then prescribe various limitations on emissions required to meet these ambient air quality standards, and individual states take appropriate actions to limit overall emission levels to comply. The emission caps and permit restrictions represent a finite level of pollution permitted for a range of industrial activities in a defined area—including electricity production.

Naturally, the permissible levels of emissions have decreased over time, as restrictions have become tighter—while the total amount of electricity needed to satisfy demand has increased. Much of the burden for reducing concentrations of harmful air pollutants to meet Clean Air Act requirements has been focused on the electric utility industry because of the ease and cost-effectiveness of controlling large, stationary sources of emissions.

But reducing emissions is not the only method employed to achieve compliance with increasingly stringent Clean Air Act limitations. Avoiding the emissions in the first place—while meeting increased electricity demand—has also been an important compliance tool. For example, between 1970 and 1990, the increased use of nuclear energy eliminated more nitrogen oxide emissions than all other actions taken to comply with Clean Air Act requirements. Nuclear energy, by avoiding additional emissions as electricity output grows, acts as a vital partner in Clean Air Act compliance.

So at the same time the United States was responding to the oil and gas shocks of the 1970s by re-balancing its energy supply portfolio to include nuclear energy, it was also aiding in the implementation of Clean Air Act requirements in states where nuclear plants operated.

The Clean Air Act doesn't just impact the power generation business. All industrial activities that emit controlled pollutants are captured under the act's provisions. That means that a state out of compliance with Clean Air Act regulations is constrained when it comes to building new conventional power plants as well as other industrial and manufacturing facilities—the building blocks of economic development. States are even suing one another over the transport of air pollution because such pollution has both environmental and economic consequences. States simply object to paying twice for another state's pollution—first in terms of the cost of cleanup caused by transported air emissions and again because of opportunities lost because the polluting state's added air emissions constrained economic development in the downwind state.

How does this pertain to nuclear energy? It pertains to nuclear energy because the high technology investment in environmental integrity that nuclear power plants represent incorporates the cost of environmental compliance from the outset. In other words, the entire environmental cost of generating electricity with nuclear power—and every generation type has an environmental cost—is, and always has been, factored into the cost of doing business.

By making a greater investment in fission technology that avoided harmful air pollutants, by internalizing the cost of safely storing, monitoring and managing the small volume of solid waste generated by nuclear plants in the form of used nuclear fuel, and by pre-funding plant decommissioning and site restoration requirements, and providing for liability insurance—the nuclear industry has made an investment in the environment and the future.

In this regard, it is important to note that the commercial nuclear industry's waste management efforts over the past 40 years have worked so well that they have set a standard that other industries seek to match as they struggle to manage disposal and clean-up of hazardous materials. I ask you: What other industry, what other technology, can honestly say that it knows precisely where—and in what amounts—its entire inventory of hazardous waste is safely located? Only the commercial nuclear energy industry.

But my point remains the same: By internalizing its external environmental costs, the nuclear energy industry made a market savvy investment—an investment that merits recognition and participation in market-based plans designed to better the environment. This is particularly true when it comes to international efforts to limit greenhouse gas emissions to forestall the possibly negative effect of global climate change. As the chairman of the Intergovernmental Panel on Climate Change Robert Watson has said:

> You have to internalize the environmental externalities into the price of energy. ...It is bad economics not to internalize an externality when that is the true social cost.

That is precisely what the U.S. nuclear energy industry has done: It has internalized the environmental externalities into the price of energy.

Lessons learned about the role nuclear energy plays in meeting Clean Air Act requirements during the last 20 years are part of the long-term, technology-based solutions that will be needed to control man-made greenhouse gases—such as carbon dioxide or methane.

As with pollutants controlled under the Clean Air Act, climate change policies generally focus on sources that emit greenhouse gases or on technologies that reduce them. This narrow view fails to recognize the embedded reliance on non-emitting technology for emissions control. When carbon reduction targets were set, nuclear energy's generation—and its emission free nature—were simply assumed as part of the mix.

According to the IAEA, nuclear power annually avoids 8 percent of the global emission of carbon dioxide from energy production. Not only would global carbon dioxide emissions be that much higher without that embedded nuclear energy emission avoidance, the only suitable replacement for that lost generation are large-scale fossil plants that produce significant amounts of additional greenhouse gases.

That is why NEI believes firmly that a ton of carbon avoided is as valuable as a ton of carbon reduced. That is why NEI believes that efforts to exclude nuclear energy from international credit trading and technology transfer programs under the Kyoto Protocol's Clean Development Mechanism are arbitrary and discriminatory. And that is why NEI believes denying nuclear energy access to incentives under the Kyoto Protocol undermines the credibility of the international effort—and could actually foster an increase in greenhouse gas emissions. That is why the European Commissioner for Energy Loyola de Palacio recently said that if Europe decommissions its nuclear units, it could forget about meeting the greenhouse gas emission reductions envisioned in the Kyoto Protocol.

As an American sports legend once remarked, there is nothing so hard to predict as the future. But we can be fairly certain about some broad trends.

First, the world's population will continue to grow. One estimate is by more than 50 percent by 2050. Second, that population will increasingly be concentrated in intensive energy-consuming urban areas. Third, as the world's population grows, so will its appetite and need for energy—and electricity in particular. Otherwise, continued social and economic development on a global scale will be impossible.

Is there any one energy source that can meet rising demand? Of course not. All sources of electricity—nuclear, fossil, hydro and renewables—are needed to power and protect the future. Are some sources more suitable than others are for certain applications? Absolutely. There is no either/or option. The only option is to develop—in a thoughtful manner—the various energy sources that, in combination, can work together for the future. Nuclear energy must continue to be a part of that mix—and it must be part of the future.

Thank you.

INDEX

MIX
Papier aus verantwortungsvollen Quellen
Paper from responsible sources
FSC® C105338

If you have any concerns about our products,
you can contact us on
ProductSafety@springernature.com

In case Publisher is established outside the EU,
the EU authorized representative is:
Springer Nature Customer Service Center GmbH
Europaplatz 3, 69115 Heidelberg, Germany

Printed by Libri Plureos GmbH
in Hamburg, Germany